The
Mature
Mind

ALSO BY GENE D. COHEN, M.D., PH.D.

The Creative Age: Awakening Human Potential
in the Second Half of Life

THE MATURE MIND

The Positive Power of the Aging Brain

GENE D. COHEN, M.D., PH.D.

A Member of the Perseus Books Group
New York

Designed by Deborah Gayle

Library of Congress Cataloging-in-Publication Data

Cohen, Gene D.
  The mature mind : the positive power of the aging brain / by Gene Cohen.
    p. cm.
  Includes bibliographical references and index.
  ISBN-13: 978-0-465-01203-9 (hardcover : alk. paper)
  ISBN-10: 0-465-01203-5 (hardcover : alk. paper)
  1. Middle age—Psychological aspects. 2. Middle-aged persons—Psychology. 3.
Old age—Psychological aspects. 4. Older persons—Psychology. 5. Aging—Psycho-
logical aspects. 6. Brain—Aging. I. Title.
BF724.6.C64 2005
155.67'13—dc22

                                                                    2005022508

05 06 07 08 / 10 9 8 7 6 5 4 3 2 1

To the elders of my family,
in deep appreciation
for the nurturing and wisdom
they have provided to our family
and to the community.

Contents

Acknowledgments

FIRST, I WOULD LIKE TO ACKNOWLEDGE the great support and patience my family provided me while I wrote this book. The fact that for ten months of the process I was recovering from a broken femur and required two operations made their support all the more precious. My wife, Wendy Miller, and my daughter, Eliana, were simply extraordinary during this time, and my son, Alex, his wife, Kate, and my two granddaughters—Ruby and Lucy—were a fabulous cheering section from their home in Camden, Maine.

I also want to acknowledge the invaluable advice, assistance, and encouragement that Teresa Barker provided in the planning, initial drafting, and development of the book. I am also enormously grateful to Stephen Braun, who collaborated very closely with me in bringing the final draft to fruition, contributing vitally to its quality and timeliness.

Much gratitude goes as well to my agent, Gail Ross, for her successful efforts in connecting me with Basic Books and for her ongoing assistance in keeping things on course. And many thanks to Howard Yoon, who worked so diligently with Gail in this process.

Thanks are due, too, to Jo Ann Miller, my editor at Basic Books, who worked closely with me, and was always prompt and helpful in answering questions, solving problems, and moving things along.

Deep appreciation goes to the sponsors of the major studies I describe in this book. The Atlantic Philanthropies (USA) Inc. supported the 21st Century Retirement Study. In conjunction with the Helen Bader Foundation, they also supported the development and evaluation of the reading list of children's books described in Appendix 2.

The Creativity and Aging Study was supported by six sponsors led by the National Endowment for the Arts (NEA). The other five supporters were the Center for Mental Health Services of the U.S. Department of Health and Human Services, the National Institute of Mental Health at the National Institutes of Health, AARP, the Stella and Charles Guttman Foundation, and the International Foundation for Music Research. Particular thanks for this study go to Paula Terry, the project officer at NEA. After my book *The Creative Age* came out, Paula, who heads the Accessibility Office Program and coordinates projects on aging at the National Endowment for the Arts, read it and was interested in the summary of research relevant to the impact of creative expression on health in later life. The Arts Endowment has long been committed to making the full spectrum of the arts available to underserved populations, including older adults. Realizing that little data addressed the impact of professional arts programming on older adults, Paula encouraged me to develop guidelines for a study and to submit a proposal to the Endowment.

Finally, I want to express my special appreciation to my colleagues at the three research sites of the Creativity and Aging

Study that I directed. Jeanne Kelly, from the Levine School of Music, served as the artistic director for the metropolitan Washington, D.C., part of the study. Jeff Chapline, who heads the Center for Elders and Youth in the Arts (CEYA), directed the San Francisco site. Susan Perlstein, who heads Elders Share the Arts (ESTA), directed the Brooklyn site and shared the important findings of the Creativity and Aging Study with the National Center for Creative Aging (NCCA), which she also directs, to promote dissemination of the research results for practical use by community-based art programs across the country. Working with these terrific colleagues was like being on a dream team.

Introduction

The greatest obstacle to discovery is not ignorance—
it is the illusion of knowledge.
                                    —Daniel J. Boorstin

"Over the hill."

"Out to pasture."

"Twilight years."

"Retired."

These words reflect a stubborn myth—that aging is a negative experience and that "successful aging" amounts to nothing more than slowing the inevitable decline of body and mind. Rubbish. Some of life's most precious gifts can *only* be acquired with age: wisdom, for example, and mastery in hundreds of different spheres of human experience that requires decades of learning. Growing old can be filled with positive experiences, and "successful" aging means harnessing and manifesting the enormous positive potential that each one of us has for growth, love, and happiness.

Of course, aging brings challenges and losses. As actress Bette Davis once famously quipped, "Getting old isn't for sissies." Sight may blur, hearing may dull, friends may die or become disabled. All of this is true, but it's not the *whole* truth. Historically, both science and culture in Western societies have focused exclusively on the negative sides of aging and ignored the positive. It's time for a better, truer, and more motivating paradigm—not a rosy, everything-is-wonderful perspective, but a clear-eyed view that acknowledges the hard realities of growing old while at the same time celebrating its benefits, pleasures, and rewards. With this book I want to shatter the illusions of "knowledge" about aging that are based on faulty reasoning, insufficient research, and a preoccupation with disease and pathology. My picture of positive aging is based on cutting-edge scientific research as well as my personal experience as a psychiatrist who has treated older adults and their families for more than thirty-five years.

The latest research findings are encouraging and important. Denying or trivializing the positive potential of aging prevents people from realizing the full spectrum of their talents, intelligence, and emotions. But when we come instead to *expect* positive growth with age, such growth can be nurtured. We are still a long way from fully realizing this shift in perspective, but I hope this book will be a forceful catalyst for change in that direction.

Introduction

New Science, New Horizons

Some of the most exciting research supporting the concept of positive aging comes from recent studies of the brain and mind. Much of aging research conducted during the twentieth century emphasized improving the health of the aging body. As a result of this research, life expectancy and overall health did in fact improve dramatically. Aging research at the beginning of the twenty-first century, in contrast, has expanded with a strong focus on improving the health of the aging *mind*. Dozens of new findings are overturning the notion that "you can't teach old dogs new tricks." It turns out that not only can old dogs learn well, they are actually better at many types of intellectual tasks than young dogs.

The big news is that the brain is far more flexible and adaptable than once thought. Not only does the brain retain its capacity to form new memories, which entails making new connections between brain cells, but it can grow entirely new brain cells—a stunning finding filled with potential. We've also learned that older brains can process information in a dramatically different way than younger brains. Older people can use both sides of their brains for tasks that younger people use only one side to accomplish. A great deal of scientific work has also confirmed the "use it or lose it" adage: the mind grows stronger from use and from being challenged in the same way that muscles grow stronger from exercise.

But the brain isn't the only part of ourselves with more potential than we thought. Our personalities, creativity, and psychological "selves" continue to develop throughout life. This might sound obvious, but for many decades scientists who study human behavior did not share this view. In fact, until late in the twentieth century, psychological development in the second half of life attracted little scientific attention, and when attention was paid, often the wrong conclusions were drawn. For example, Sigmund Freud, whose influence on psychological theory was profound, had this to say about older adults: "About the age of fifty, the elasticity of the mental processes on which treatment depends is, as a rule, lacking. Old people are no longer educable."

Ironically, Freud wrote this statement in 1907, when he was fifty-one, and he wrote some of his greatest works after the age of sixty-five. Furthermore, Sophocles's *Oedipus Rex*, the masterpiece on which Freud based his pioneering psychoanalytic theory, was written when the Greek playwright was seventy-one years old.

Freud wasn't the only pioneer to get things wrong when it came to aging. Jean Piaget, who made an extraordinary contribution to our understanding of cognitive development, ended his description of intellectual development with what he called "formal operations," the kind of abstract thinking that matures during the teenage years. As far as Piaget was concerned, development stopped in young adulthood and then began a slow erosion.

Even the great developmental psychologist Erik Erikson, one of my teachers at Harvard, gave only limited attention to development in older age. Erikson delineated eight stages of psychosocial development and defined each one in terms of an issue or conflict that must be resolved. Yet only one of his stages refers to development after the onset of adulthood—mature age, which, these days, amounts to a single stage that can last fifty years! His classic work *Identity and the Life Cycle* included only one page on each of the two last stages of human life. To his credit, Erikson was one of the first influential thinkers to assert that some kind of psychosocial development continues throughout the life cycle. He acknowledged that his work on aging was incomplete, and he invited his students to continue in this area. This book is, in part, my response to the challenge Erikson made decades ago.

Four Phases

In this book I present a new account of psychological development in the second half of life. This new view explains many things about older age and is fundamentally forward-thinking and optimistic about our potential for lifelong growth, creativity, and emotional fulfillment. Based on my studies of more than 3,000 older adults, using in-depth interviews and questionnaires conducted multiple times over the years, I have identified the following four distinct developmental phases of late life: midlife reevaluation, liberation, summing up, and encore.

People enter and pass through these phases under the impelling force of inner drives, desires, and urges that wax and wane throughout life. I call these drives the "Inner Push" and have witnessed it in thousands of older adults who have participated in my research projects and clinical practice. The Inner Push is the fuel motivating development; it works in concert with the changes in the aging brain that I explore in chapter 1. My conception of phases is more fluid and dynamic than Erikson's stages because I recognize that by later life people vary widely in every conceivable way, and no rigid system will be accurate for everyone. The phases

I propose are real—I've seen them manifested time after time—but people experience them in different ways and sometimes in a slightly different sequence than the one I present.

The first phase, midlife reevaluation, is a time for exploration and transition. It is not at all the same thing as a "midlife crisis"—which modern research has shown has been overreported and is largely a cultural myth. Only 10 percent of people I interviewed reported having a midlife crisis. What I found, instead, was that in this period, from roughly ages forty through sixty-five, people undergo a profound reevaluation, asking themselves: Where have I been? Where am I now? Where am I going? Most people experience this period not as a crisis but as a quest—a desire to break new ground, answer deep questions, and search for what is true and meaningful in their lives.

The midlife reevaluation phase is followed by what I call the liberation phase: a time when we feel a desire to experiment, innovate, and free ourselves from earlier inhibitions or limitations. This desire often overlaps with midlife reevaluation and then comes on strong throughout the late fifties and sixties and into the seventies. As this shift is happening, our brains undergo significant physiological changes, including the sprouting of new connections between brain cells and a more balanced use of the two brain hemispheres. This is a time when people express the sense of "If not now, when?"

The summing up phase, from the late sixties through the seventies and eighties, can be a time of recapitulation, resolution, and review. One of the common outcomes of this autobiographical summing up process is a desire to give back—to family, friends, and society. Volunteerism and philanthropy, prominent among older people well into their eighties, are two tangible manifestations of this phase.

For the final phase I use "encore" in the French sense of "again," "still," and "continuing." This phase is not a swan song, but a variation on a theme: the desire to go on, even in the face of adversity or loss. This need to remain vital can lead to new manifestations of creativity and social engagement that make this period full of surprises.

When people come to understand these phases of later life and the mechanisms at work behind them, I have seen them become powerfully motivated and energized. Released from overly negative illusions about aging, people are often stirred by new energy, direction, or purpose.

DEVELOPMENTAL INTELLIGENCE

In this book I introduce a novel concept, developmental intelligence, which I see as the greatest benefit of the aging brain/mind. Developmental intelligence is the degree to which a person has manifested his or her unique neurological, emotional, intellectual, and psychological capacities. It is also the process by which these elements become optimally integrated in the mature brain. More specifically, developmental intelligence reflects the maturing synergy of cognition, emotional intelligence, judgment, social skills, life experience, and consciousness. We are all developmentally intelligent to one degree or another, and, as with all intelligence, we can actively promote its growth. As we mature, developmental intelligence is expressed in deepening wisdom, judgment, perspec-

tive, and vision. Advanced developmental intelligence is characterized by three types of thinking and reasoning that develop later
than Piaget's "formal operations" and hence are referred to as
"postformal operations": relativistic thinking (recognizing that
knowledge may be relative and not absolute); dualistic thinking
(the ability to uncover and resolve contradictions in opposing and
seemingly incompatible views); and systematic thinking (being
able to see the larger picture, to distinguish between the forest and
the trees).

These three types of thinking are "advanced" in the sense that
they do not come naturally in youth; we prefer our answers black
or white, right or wrong. And we usually prefer *any* answer to none
at all. It takes time, experience, and effort to develop more flexible
and subtle thinking. Our capacity to accept uncertainty, to admit
that answers *are* often relative, and to suspend judgment for a more
careful evaluation of opposing claims is a true measure of our
developmental intelligence. In this book I'll show you how you can
cultivate your developmental intelligence and thereby reap its
rewards.

Two New Studies

I've had the privilege of directing two groundbreaking studies of
older age since 2000, one looking at the new face of retirement and
the other at the positive benefits of creativity in older adults. Both
studies have generated surprising—and encouraging—results. My
retirement study shows just how outmoded the word "retirement"
really is. For most people these days, the years after age sixty-five

are anything but "retiring." It's not that everyone is a dervish or that people don't relax and enjoy themselves, but most people I interview see this life stage as a great opportunity to pursue activities and interests for which they previously didn't have time. Far from being a time of social and cultural withdrawal (as was postulated by early influential research), "retirement" can usher in *greater* engagement, more satisfying relationships, new intellectual growth, and more fun.

My other study explores the mental, physical, and emotional effects of participating in a community arts program. Again, my colleagues and I have made surprising discoveries. Contrary to societal myths, creativity is hardly the exclusive province of youth. It can blossom at any age—and in fact it can bloom with more depth and richness in older adults because it is informed by their vast stores of knowledge and experience. As I will explain more fully later, taking part in any kind of art program, including the nonvisual arts of music, dance, and theater, can improve your health, your outlook, and your resilience.

Important implications flow from these two studies, both for private citizens and for those responsible for supporting the health and well-being of older adults. Outcomes from our creativity study, for example, should be invaluable for program directors of senior centers. Similarly, the finding from the retirement study that many older people are seeking part-time work should interest human resources directors in corporations and nonprofit institutions. In reporting the findings from my research, I hope to provide a road map for improving the social supports and educational opportunities for all older adults.

MY HOPE

In 1971, when I entered the field of gerontology, it was a relatively new area of study, underfunded and hobbled by stereotypes and misconceptions. As recently as the 1960s and 1970s, many experts still viewed old age as a disease: the mind and body, they believed, naturally fell apart, like a car after many years of use.

By the mid-1970s, these views began to change as evidence accumulated about the realities of aging and as the population of older adults began to surge. The federal government started spending millions of dollars on new research through two major programs: the National Institute on Aging and the Center for Studies of the Mental Health of the Aging, the latter of which I was fortunate enough to be the first director. Researchers began to understand that aging is not a medical condition in and of itself; it is simply a time of life in which many medical conditions become manifest—the so-called age-associated problems. This new focus fueled the field of geriatrics and provided a more balanced view of old age. Healthy adults, researchers found, retain sound mental and emotional faculties and typically decline only gradually in their physical resources.

Over the next thirty years, funding for aging research grew from $50 million a year to more than $1 billion today. Yet despite this infusion of time and money, studies still tend to focus on the problems of older age. Even the recent and important book *Successful Aging*, by John Rowe and Robert Kahn, presents the goal as minimizing decline rather than recognizing the huge potential for positive growth in later life. Although Rowe and Kahn rightly emphasize the importance of maintaining health, mental functioning, and active engagement in life, they don't present the possibilities for *improving* these areas with age.

The Mature Mind presents a new paradigm of aging, one that I hope will eventually displace today's negative views and assumptions. It recognizes the potential beyond the problems associated with aging. It reframes the aging process as a set of developmental phases that support real growth as opposed to the view of aging as an inevitable decline. This book shows how we can support and cultivate our natural capacity for positive change. I sincerely hope it will help redirect public dialogue on this topic by delivering a promising message about the value and capacities of the maturing mind.

1

The Power of Older Minds

Your brain never stops developing and changing.
It's been doing it from the time you were an embryo,
and will keep on doing it all your life.
And this ability, perhaps, represents its greatest strength.
—James Trefil, physicist and author

MY IN-LAWS, HOWARD AND GISELE MILLER, both in their seventies, were stuck. They had just emerged from the Washington, D.C., subway system into a driving snowstorm. They were coming to our house for dinner and needed to catch a cab because it was too far to walk—but it was rush hour, and no cabs stopped. Howard tried calling us to get a lift, but my wife and I were both tied up in traffic and weren't home yet.

As his fingers began to turn numb from the cold, Howard noticed the steamy windows of a pizza shop across the street. He and Gisele marched through the slush, entered the shop, stepped up to the counter, and ordered a large pizza for delivery. When the cashier asked where to deliver it, Howard gave him our address, and added, "Oh, there's one more thing."

"What's that?" the cashier asked.

"We want you to deliver us with it," Howard said.

And that's how they arrived—pizza in hand—for dinner that night.

This favorite family story perfectly illustrates the sort of agile creativity that the aging mind can produce. Would a younger person have thought of this solution? Possibly. Creativity knows no age limits. But in my experience, this kind of out-of-the-box thinking is a learned trait that improves with age. Sherry Willis, of the Human Development and Family Studies program at Pennsylvania State University, calls it pragmatic creativity in everyday problem solving, a capacity that her research has found to be very strong in later life. Age allows our brains to accumulate a repertoire of strategies developed from a lifetime of experience—part of what has been referred to by other researchers as crystallized intelligence. Howard hadn't done the pizza parlor routine before, but the accumulated experience of other successful strategies helped stimulate the thinking that produced his creative solution.

Howard's solution reflects not only the experience of years and a certain agility of thought but also a mature psychological development that is prevalent among people in their sixties and seventies. With age can come a new feeling of inner freedom, self-confidence, and liberation from social constraints that allows for novel or bold behavior. Howard wasn't afraid to make an unusual request of perfect strangers, and that was a key part of his success that night.

The Power of Older Minds

In Howard's story we have a picture of a healthy aging mind at work: clear, creative, resourceful, and powerful. But how does such a mind develop? On what does it depend for its existence? The short answer is the brain.

It's been said that the mind is what the brain does. The mind is often described as "software" running on the "hardware" of the brain. But this analogy is too simple. The brain is far more malleable and flexible than any computer chip. And the mind, although it seems almost ghostlike, can powerfully influence the brain and, by extension, the body. Mind and brain are really two sides of a single coin—mind/brain. This chapter explores the brain side of this equation and looks at recent discoveries in brain science that illuminate the positive potential of the aging mind.

You may have learned the following "facts" about the brain:

- The brain cannot grow new brain cells.
- Older adults can't learn as well as young people.
- Connections between neurons are relatively fixed throughout life.
- Intelligence is a matter of how many neurons you have and how fast those neurons work.

All these "facts" are wrong, as we will see. And that's good news for all of us. The brain is more resilient, adaptable, and capable than we long thought. Research in the past two decades has established four key attributes of the brain that lay the foundation for an optimistic view of human potential in the second half of life:

- The brain is continually resculpting itself in response to experience and learning.
- New brain cells *do* form throughout life.
- The brain's emotional circuitry matures and becomes more balanced with age.
- The brain's two hemispheres are more equally used by older adults.

Now, let's be clear. I am not suggesting that the brain is immune to age-related changes. The brain is made of cells, like every other part of the body, and cells can and do "wear out" with age. Certain aspects of brain function do decline with age, such as the raw speed with which complicated math problems are solved, reaction times, and the efficiency of short-term memory storage. But these "negatives" are by no means the whole—or even the most important—story about the aging brain. Unfortunately, because much brain research has focused on age-related *problems*, negative aspects of aging have been emphasized and the positive implications of research have been overlooked. Indeed, one of the most important findings of all is still not widely known; namely, that much of the decline in mental abilities formerly associated with aging is *not* caused by aging per se but by specific diseases such as "microstrokes," Alzheimer's disease, and mental illnesses such as depression. Healthy older brains are often as good as or better than younger brains in a wide variety of tasks.

Understanding more about how your brain works is important because understanding can spur motivation. If you learn how memory works and see the connections between the health of your neurons and the choices you make in diet, exercise, sleep, social activity, and how you challenge your mind, you'll be more likely to harness your brain's latent potential.

THE POTENTIAL OF OLDER BRAINS

The most important difference between older brains and younger brains is also the easiest to overlook: older brains have learned more than younger brains. Many aspects of life are simply too complicated and subtle to learn quickly, which is why experience counts in so many spheres of life. Human relationships, for example, are notoriously complicated, and it can take decades to acquire the deep knowledge and understanding it takes to be a truly effective therapist, pastor, manager, or politician. There is simply no substitute for acquired learning in such fields as editing, law, medicine, coaching, and many areas of science. In these and many others fields, age generally trumps youth. Of course, age alone is no guarantee of excellence, but excellence in many fields can be achieved only after many years of hard work and experience.

We now know that learning actually causes physical changes in the brain. An older adult's brain, magnified tremendously, would look distinctly different from a young person's brain. The brain cells (called neurons) in the parts of the brain that an older person has used continuously would look like a dense forest of thickly branched trees, compared with the thinner and less dense forest of a young brain. This neural density is the physical basis for the skills of accomplished older adults.

Let's take a closer look at learning and how it sculpts the brain.

To learn, we must remember. Memories, in turn, are created when clusters of hundreds or thousands of neurons fire in a unique pattern. Whenever you perceive anything, whether it's a whiff of cinnamon, a catchy song, or a visual image, a flood of signals lights up particular constellations of neurons in certain parts of your brain. If conditions are right (that is, if you are paying attention), the connections between these neurons are automatically strengthened. If this particular pattern of neurons is stimulated later in exactly the same way—say, by another whiff of cinnamon—the network "lights up" more easily than it did previously and you "remember" the smell. The original sensation is stored in these discrete patterns of primed connections. The more often a particular pattern is stimulated, the more sensitive and permanent are the connections between the neurons in the pattern. This process of memory formation is summarized by the phrase "neurons that fire together, wire together."

Not only does learning link neurons in new patterns, it also stimulates neurons to grow *new* connections (known as synapses) through tiny branchlike extensions called dendrites.

The idea that the brain physically changes as a result of learning—a phenomenon called plasticity—emerged in the mid-1960s from animal studies conducted by Marion Diamond, a professor of anatomy at the University of California, Berkeley. Diamond found that when rats were placed in a more stimulating environment, their neurons sprouted new dendrites and produced higher levels of an important brain chemical called acetylcholine. The age of the rats made no difference. The brains of older rats showed the same kind of robust response to a stimulating environment as the brains of younger rats. A great deal of research conducted since those pioneer-

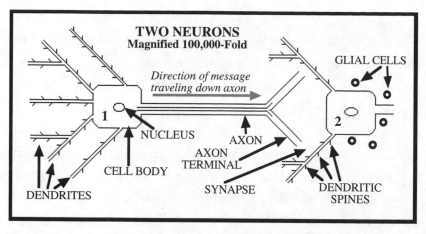

Two neurons magnified 100,000-fold showing dendrites, cell body, axon, and synapse(s) with postsynaptic neuron(s). Regardless of age, a stimulating environment with challenging activities prompts an increased production of dendrites, dendritic spines, glial cells, and synapses (contact points between different neurons), enhancing brain capacity and communication among neurons.

ing experiments has proven that the same phenomenon occurs in all animals, including human beings. In short, the brain actively grows and rewires itself in response to stimulation and learning.

For example, brain imaging studies have found that London taxi drivers have an enlarged region in the hippocampus—the part of the brain used for navigating in three-dimensional space. The drivers' experience of negotiating the complex London streets "exercised" that part of their brains, which grew as a result. (And the oldest drivers in the study showed no hippocampal decline compared with the younger drivers.) Similar findings were reported in a study of musicians, who were found to have significantly increased functioning in the parts of the brain associated with hearing and the discrimination of tone and pitch.

I suspect that a brain scan of my neighbor, Lorraine Kennedy, would reveal similar kinds of robust growth in certain parts of her brain. Kennedy, who just turned ninety, is legendary for her knowledge of local history. She is an infallible resource on who lived where, and when, in the Victorian houses lining our street. Whenever some clarification is needed about a historical fact, people turn to her, and, without a computer, digital assistant, or notes, she retrieves the information from memory—sometimes along with plenty of additional detail. The part of her brain where all this knowledge resides would look far richer and more complicated than the corresponding part in the brain of a young person.

The complex neural architecture of older brains, built over years of experience, practice, and daily living, is a fundamental strength of older adults. And the more complex the architecture, the more it resists degradation by injury or disease.

Of course, the brain and its architecture are not static. Our brains are a bit like Monticello, the home that Thomas Jefferson built and then proceeded to modify during the many years he lived there. By continuing to learn and have new experiences, we can actively maintain, build, and remodel our brains for more effective and creative tasks. Doing so involves avoiding certain things as well. Stress, excessive alcohol and drug use, inactivity, smoking, obesity, malnourishment, and social isolation all weaken the brain's neural superstructure. In fact, these are some of the real culprits behind age-related mental decline, not aging itself.

One aspect of brain development is frequently misunderstood, to the detriment of people's views about their own mental capacities. Some people have the false belief that mental ability is "all in the genes" and that anything that has a genetically determined component, such as eye color or the shape of one's nose, can't be changed. Genes, of course, are very powerful, and, broadly speaking, the genes we inherit do set some limits on what we can achieve mentally or physically. No matter how hard some people train, they'll simply never be Olympic runners, for example. But genes are not all-powerful—not by a long shot. Rather than constituting an immutable blueprint for our bodies and behavior, genes, it turns out, are highly sensitive to our environment, what we are exposed to, what we perceive, the emotions we feel, the stress we are under, and a host of other factors in our lives. Many genes include "switches" for their activity—they can be turned on or off, or their activity levels can be dialed up or down like a volume control.

Thousands of genes are involved in the growth and maintenance of the brain. In fact, the brain uses up a larger portion of the human genome than any other organ. Many of these genes respond to the neural stimulation of learning. By challenging our brains, we not only actively shape existing neurons and stimulate their growth, we also switch on the key genes for making the cellular raw materials needed for mental development. The relationship between us and our genes is much more like a dance than most people think, which is a hopeful situation indeed.

ESTELLE'S STORY

The potential of the healthy older brain is beautifully epitomized in Estelle Jansen, age seventy-one, a participant in my study of retirement-age adults. Estelle lives independently in a retirement community, where she moved after her husband's death.

⸙Estelle's life has been rich and full. Her husband, to whom she was married for forty-one years, was a foreign service officer, and his career took them to many countries around the world. Estelle loved the experience of living abroad and always tried to learn the local language wherever they landed.

After recovering from her husband's death, Estelle began again to feel the desire for fresh, challenging experiences. She decided to earn a master's degree in history, despite some reservations about not having computer skills and fears about whether she could keep up with classmates in their twenties. But she dove into the challenge, starting off by taking a computer course for older adults who were new to the technology. She said it was "like learning a foreign language," and she quickly learned what she needed for her course work.

In her history courses, she found that her younger colleagues enjoyed her contributions to class discussions. And because of her focus on the work—including her habit of spending more time on assignments by choice, rather than need—she was very successful in her classes. When I met her, she had just completed her first semester and earned solid grades. "I kept up just fine after all, thank you!" she said.

You might think that Estelle's brain, so able and open to learning at seventy, is exceptional. But in fact all older adults, unless they are afflicted with injury or disease, have the same capacity to learn, grow, and derive satisfaction and pleasure from new accomplishments. Indeed, the fastest-growing group of graduate students are those over age fifty, a trend that supports an optimistic view of the potential of the aging brain.

⁎New Brain Cells, New Potential

For decades, one of the most unshakable axioms of neuroscience was that nerve cells cannot regrow and the brain does not create new neurons. Then, in the early 1960s, Joseph Altman, at the Massachusetts Institute of Technology, made the startling discovery that new neurons *can* form—at least in rats. Specifically, he found new cells growing in the hippocampus of adult rats, a part of the brain that is critical for new memory formation. In 1998, scientists showed that the adult human brain produces new neurons as well, a process called neurogenesis. It turns out that many regions of the brain have primitive cells that, under certain conditions, can mature into either fully functioning neurons or brain cells called glia, which provide mechanical and nutritional support for the neurons. (It's probably not coincidental that, at autopsy, Einstein's brain was found to have significantly more glial cells than average.)

These remarkable findings were rich with implications, not only for the treatment of brain-decaying diseases such as Parkinson's disease and Alzheimer's disease, but for anyone interested in preserving or improving their mental capacity. The findings threw open the possibility that we can actually build (or rebuild) our brains.

We now know that new brain cells can form in other important regions of the brain. In 1999, a team of scientists at Princeton University led by psychologist Elizabeth Gould found that, in monkeys, new neurons can grow in several regions of the cerebral cortex, the region responsible in human beings for many of our "higher" functions, such as reflection, planning, decision-making, and emotional control.

Researchers are beginning to home in on what triggers new neuronal growth, although much about the process remains unknown. Challenging mental activity certainly stimulates neuronal growth, but so does another activity that few would have suspected: vigorous physical exercise. Exercise seems to "juice" the brain by stimulating the production of chemicals called brain growth factors. These compounds, in turn, provoke the primitive brain cells to mature into neurons. Prolonged stress, on the other hand, seems to dramatically suppress new neuron production.[i] Studies have shown that both physical and psychological stress reduce the growth of new cells in the hippocampus. Patients with depression or posttraumatic stress disorder also show reduced hippocampal volume, while treatment reverses this trend.

The discovery that new brain cell growth can occur in adulthood has transformed views of the aging brain and the potential for enhancing brain functioning in older adults. For example, Gerd Kempermann, of the Department of Experimental Neurology at Humboldt University in Berlin, says that studies conducted in his department suggest that neurogenesis enables the aging brain "to accommodate continued bouts of novelty." In other words, neurogenesis may have played a role in my father-in-law's pizza delivery inspiration[i] and Estelle Jansen's success in a graduate program at age seventy.

Fred Gage, at the Laboratory of Genetics at the Salk Institute, who is one of the co-discoverers of neurogenesis in the mature brain, notes that "the stage has now been set for a new understanding of the adult brain" and emphasizes that even a damaged adult brain very likely retains some capacity to heal and recover. Of course, since memories are stored in patterns of connection between existing neurons, new neurons will not mean the recovery of memories lost to disease or injury. But the brain's ability to grow new neurons is one of the most exciting discoveries in neuroscience and a dramatic reason for optimism about the brain's potential in the second half of life.

EMOTIONS IN BALANCE

George Barker was a smart, scrappy son of a housepainter. When World War II broke out, he enlisted in the Air Force and became a bomber pilot. Like many veterans, the experience of war and the real possibility of being killed affected him deeply—although in the decades following his combat experience he rarely talked about it. He returned from the front, began a career in journalism, married, and raised a family. Then, when he was in his sixties, he was diagnosed with leukemia, a cancer of white blood cells that weakens the immune system.

His doctors advised him to avoid travel and any interactions that would expose him to pathogens, such as playing with his grandchildren. He rejected the advice.

"I went through a war being shot at every day," he says. "I'm not going to start living in fear now because I might catch a cold."

Taking some appropriate cautions, he continued to travel, see his family, and engage in a full and loving life. A disease that might have caused a younger person to become depressed or withdrawn became a challenge, but not an obstacle.

Among the myriad negative stereotypes and myths about aging is the view that many older people are depressed by the physical ills that, admittedly, become more common with age. In fact, however, research shows that the incidence of depression in later life is no higher than in early adulthood. Many studies have reported high morale in older adults, even among those who are frail. The positive outlook of people like Estelle Jansen and George Barker isn't the exception—it's the rule. I recall one day saying, "Have a lovely day!" to a seventy-four-year-old woman participating in one of my studies.

"I'll *make* it a lovely day!" she shot back with a smile.

Many factors play into the high morale and positive outlook of so many older persons, including such things as a greater acceptance of life's realities, a greater sense of self, and a long-term perspective that makes it easier to accept the inevitable slings and arrows of daily life. In addition to such psychological factors, however, new evidence shows that changes in the older brain itself play an important role in the emotional aplomb and equanimity of many older adults.

Human emotional responses are produced and regulated by a set of structures deep in the brain called the limbic system. These structures have been shaped by millions of years of natural selection to provide both a carrot and a stick to guide behavior in ways that favor survival and reproduction. Positive emotions such as feelings of affection, bonding, love, pleasure, and happiness arise from electrical and chemical activity in the limbic system in response to external cues such as the proximity of potential mates;

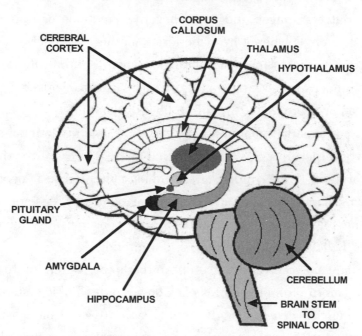

SIDE VIEW OF THE INNER BRAIN
AND THE LIMBIC SYSTEM

CORPUS CALLOSUM

CEREBRAL CORTEX

THALAMUS

HYPOTHALAMUS

PITUITARY GLAND

AMYGDALA

HIPPOCAMPUS

CEREBELLUM

BRAIN STEM TO SPINAL CORD

Side view of the inner brain and the limbic system. The limbic system is a group of anatomical structures, including the hippocampus, amygdala, parts of the cortex, and parts of the hypothalamus, that functions as the motivation and emotional center of the brain.

success in obtaining food, status, and security; and, in human beings, the satisfaction of higher drives such as curiosity and artistic or musical expression.

Negative emotions such as fear, anger, envy, disgust, and depression arise in response to events or situations that threaten our survival, well-being, or sense of fair play. Some negative emotions are inborn. Anger, for example, is a nearly universal response to the perception of unfairness against us. And some fears, such as fear of snakes, spiders, and heights, have a genetic component to them. But fear, anger, and other negative emotions can also be learned, such as when a hypochondriac mother engenders a constant fear of illness in her children, or when a chronically angry father unconsciously transmits the message that anger is an expected or appropriate reaction to life's difficulties.

A particularly important aspect of the human emotional system is how these systems are connected to the neocortex, the thick layer of brain tissue overlaying the limbic system. The neocortex performs many functions, among them giving us our sense of self-awareness as well as "higher" attributes of consciousness such as morals, beliefs, intentions, goals, and aspirations.

Many more nerve fibers run from the limbic system up to the cortex than run from the cortex back down to the limbic system. If neural activity were water, the limbic system would have a fire-hose connection to the cortex and a straw from the cortex. This fundamental imbalance in connectivity means that emotions can easily overwhelm and overrule the thinking, deliberating parts of our brains. From an evolutionary standpoint, this makes perfect sense: animals that respond quickly and unthinkingly to perceived threats have a much better chance of survival than animals that ponder the situation before reacting.

For human beings, however, this basic brain imbalance between our reason and our emotions leads to all sorts of trouble. Indeed, our universal tendency to be pulled between what we know is right and what we want to do is at the core of much great art and literature. Our frequent inability to control our emotions strength of the activity of their hemispheres, to ascribe things like career choices to one side of the brain or the other is a misrepresentation of current knowledge.

At the level of specific brain functions, however, we can correctly speak of right/left differences. It turns out that throughout early life the brain typically uses only one side at a time for things like decoding written language, generating speech, or recognizing patterns. Use of one side of the brain is referred to as unilateral hemisphere involvement; use of both sides is referred to as bilateral hemisphere involvement.

In the course of conducting studies with PET scans and magnetic resonance imaging, scientists noticed something unexpected going on in the brains of older adults. When, for example, young adults retrieve a specific word from memory, they usually use the left side of their brain. Older adults doing the same task, however, often use both hemispheres. This phenomenon has been found with other tasks too, such as face recognition, working memory, and certain types of perception. The part of the brain examined in these studies is the prefrontal cortex in both hemispheres, a region that lies just behind the forehead. Much of this work has been described by Roberto Cabeza, of the Center for Cognitive Neuroscience at Duke University. He calls the phenomenon "hemispheric asymmetry reduction in older adults"—dubbed HAROLD for short.

These findings were puzzling. Did the bilateral hemisphere involvement in older people reflect some kind of impairment? Was it a desperate attempt by an aging brain to draw on greater brain power to solve problems? Or was it something positive—perhaps a

and cravings is one of the defining features of our species. Our ability to control our emotions and modulate our behavior appropriately, however, is a hallmark of maturity.

The capacity to ride out emotional storms more flexibly and resiliently is one of the great fruits of aging. This is partly due to learning, experience, and practice, which stimulate the growth of new dendrites and sometimes entirely new neurons. This may mean that we can actually begin to equalize the out-of-balance connections between the limbic system and the cortex. In a real sense, we can build more control wires connecting our "higher" selves to our emotional centers.

But that's not all. The limbic system itself appears to grow calmer with age. One focus of current research is the amygdalae, two almond-shaped structures in the limbic system that generate some of our most intense emotions. The amygdalae are positioned to intercept sensory information streaming in from our eyes, ears, and noses; if that information contains a potential threat, the amygdalae immediately fire off volleys of impulses that can change our behavior even before the signals have been fully processed and interpreted by our neocortex. That's why your heart starts pounding at the vague shape of two men approaching you on a dark sidewalk. The men may or may not be a threat, but your amygdalae don't care and are preparing you for the worst.

People vary widely in the vigor with which their amygdalae respond. Brains are as unique as faces, and that goes for specific brain structures as well. Some people have very reactive, sensitive amygdalae—they startle easily, are "hot tempered," or feel intense bodily reactions to frightening situations. Others have relatively quiet amygdalae and are therefore more apt to be "cool," rational, unreactive, and unemotional. But for almost everyone, the amygdalae are notoriously difficult to control—and the younger you are, the harder they are to rein in.

This is where some recent research on the aging brain comes into play. Studies using brain imaging techniques such as positron-emission tomography (PET) scanning find that activity in the amygdalae decreases with age, specifically in response to negative emotions such as fear, anger, and hatred.

A study by Mather Canli and his colleagues in the psychology department at the University of California, Santa Cruz, found that as adults age, they:

- Experience less intense negative emotions
- Pay less attention to negative than to positive emotional stimuli
- Are less likely to remember negative than positive emotional materials

As the study's authors summed it up, "This profile of findings suggests that, with age, the amygdalae may show decreased reactivity to negative information while maintaining or increasing their reactivity to positive information."

In short, older people are usually calmer in the face of life's challenges. As one of the subjects in my retirement study put it, "I'm less up in arms about things and less of a perfectionist. Minor things don't upset me, and I make better judgments about things." We now know that this positive aspect of aging is the result not just of experience and learning but also of fundamental changes in brain function.

EXTREME MAKEOVER OF THE AGING BRA

As human beings evolved and developed the capacity to d strategies for survival, such as tool use and language, th brain increased in size. But Nature had a problem: the hea simply keep getting larger, because eventually it wouldn't fi the birth canal. The female pelvis could only enlarge so fai the entire body becoming unstable. What to do?

Conveniently, the brain, like some other organs such ears, lungs, and kidneys, is a dual structure. We actually l brains, left and right, connected by a kind of broadband ne called the corpus callosum. Nature's solution to the pro ever-increasing demands for specialized brain processing a to use a division of labor: some capacities would be hand marily by the left hemisphere, others by the right. Hence, people (there are some interesting exceptions) speech, la and mathematical and logical reasoning are handled by hemisphere. The right hemisphere tends to specialize i functions as face recognition, visual-spatial comprehensio intuitive/holistic operations, such as those underlying artist ativity.

Much has been made of these differences in hemis function, not all of it well grounded in science. Some talk o brained people" and "right-brained people" or of women more "right-brained" and men being more "left-brained." T mostly metaphor, however. Healthy men and women need sides of their brains and use both sides fluidly and continu throughout life. Although people certainly vary in the re and cravings is one of the defining features of our species. Our ability to control our emotions and modulate our behavior appropriately, however, is a hallmark of maturity.

The capacity to ride out emotional storms more flexibly and resiliently is one of the great fruits of aging. This is partly due to learning, experience, and practice, which stimulate the growth of new dendrites and sometimes entirely new neurons. This may mean that we can actually begin to equalize the out-of-balance connections between the limbic system and the cortex. In a real sense, we can build more control wires connecting our "higher" selves to our emotional centers.

But that's not all. The limbic system itself appears to grow calmer with age. One focus of current research is the amygdalae, two almond-shaped structures in the limbic system that generate some of our most intense emotions. The amygdalae are positioned to intercept sensory information streaming in from our eyes, ears, and noses; if that information contains a potential threat, the amygdalae immediately fire off volleys of impulses that can change our behavior even before the signals have been fully processed and interpreted by our neocortex. That's why your heart starts pounding at the vague shape of two men approaching you on a dark sidewalk. The men may or may not be a threat, but your amygdalae don't care and are preparing you for the worst.

People vary widely in the vigor with which their amygdalae respond. Brains are as unique as faces, and that goes for specific brain structures as well. Some people have very reactive, sensitive amygdalae—they startle easily, are "hot tempered," or feel intense bodily reactions to frightening situations. Others have relatively quiet amygdalae and are therefore more apt to be "cool," rational, unreactive, and unemotional. But for almost everyone, the amygdalae are notoriously difficult to control—and the younger you are, the harder they are to rein in.

This is where some recent research on the aging brain comes into play. Studies using brain imaging techniques such as positron-emission tomography (PET) scanning find that activity in the amygdalae decreases with age, specifically in response to negative emotions such as fear, anger, and hatred.

A study by Mather Canli and his colleagues in the psychology department at the University of California, Santa Cruz, found that as adults age, they:

- Experience less intense negative emotions
- Pay less attention to negative than to positive emotional stimuli
- Are less likely to remember negative than positive emotional materials

As the study's authors summed it up, "This profile of findings suggests that, with age, the amygdalae may show decreased reactivity to negative information while maintaining or increasing their reactivity to positive information."

In short, older people are usually calmer in the face of life's challenges. As one of the subjects in my retirement study put it, "I'm less up in arms about things and less of a perfectionist. Minor things don't upset me, and I make better judgments about things." We now know that this positive aspect of aging is the result not just of experience and learning but also of fundamental changes in brain function.

Extreme Makeover of the Aging Brain

As human beings evolved and developed the capacity to devise new strategies for survival, such as tool use and language, the human brain increased in size. But Nature had a problem: the head couldn't simply keep getting larger, because eventually it wouldn't fit through the birth canal. The female pelvis could only enlarge so far without the entire body becoming unstable. What to do?

Conveniently, the brain, like some other organs such as eyes, ears, lungs, and kidneys, is a dual structure. We actually have two brains, left and right, connected by a kind of broadband neural link called the corpus callosum. Nature's solution to the problem of ever-increasing demands for specialized brain processing areas was to use a division of labor: some capacities would be handled primarily by the left hemisphere, others by the right. Hence, in most people (there are some interesting exceptions) speech, language, and mathematical and logical reasoning are handled by the left hemisphere. The right hemisphere tends to specialize in such functions as face recognition, visual-spatial comprehension, and intuitive/holistic operations, such as those underlying artistic creativity.

Much has been made of these differences in hemispheric function, not all of it well grounded in science. Some talk of "left-brained people" and "right-brained people" or of women being more "right-brained" and men being more "left-brained." This is mostly metaphor, however. Healthy men and women need both sides of their brains and use both sides fluidly and continuously throughout life. Although people certainly vary in the relative strength of the activity of their hemispheres, to ascribe things like career choices to one side of the brain or the other is a misrepresentation of current knowledge.

At the level of specific brain functions, however, we can correctly speak of right/left differences. It turns out that throughout early life the brain typically uses only one side at a time for things like decoding written language, generating speech, or recognizing patterns. Use of one side of the brain is referred to as unilateral hemisphere involvement; use of both sides is referred to as bilateral hemisphere involvement.

In the course of conducting studies with PET scans and magnetic resonance imaging, scientists noticed something unexpected going on in the brains of older adults. When, for example, young adults retrieve a specific word from memory, they usually use the left side of their brain. Older adults doing the same task, however, often use both hemispheres. This phenomenon has been found with other tasks too, such as face recognition, working memory, and certain types of perception. The part of the brain examined in these studies is the prefrontal cortex in both hemispheres, a region that lies just behind the forehead. Much of this work has been described by Roberto Cabeza, of the Center for Cognitive Neuroscience at Duke University. He calls the phenomenon "hemispheric asymmetry reduction in older adults"—dubbed HAROLD for short.

These findings were puzzling. Did the bilateral hemisphere involvement in older people reflect some kind of impairment? Was it a desperate attempt by an aging brain to draw on greater brain power to solve problems? Or was it something positive—perhaps a way for the brain to create more "redundancy"—a backup system, so to speak?

To compare these two hypotheses, Cabeza measured brain activity during a set of memory tasks being performed by three groups: younger adults, low-performing older adults, and high-performing older adults. He found that the low-performing older adults used right prefrontal cortex regions similar to those used by the young adults, but the high-performing older adults used both hemispheres. These findings suggest that although the low-performing older adults and the young adults recruited similar networks of brain cells, the older subjects used them inefficiently (hence their low performance), whereas the high-performing older adults counteracted age-related neural decline by reorganizing their neural networks.

This is stunning news. Although we don't yet understand exactly how the brains of older adults remodel themselves in this more efficient and powerful way, they clearly do. This phenomenon may be related to another one of older age: an interest in making sense of one's life by writing and talking about it.

CHARLES'S STORY

Charles Pugh, now ninety-two, was seventy-two when he retired from a long career as an IRS agent. He moved easily into a less pressured life that included a lot of the things he'd loved but never had time for: fishing, reading, bowling with friends, and more relaxed time with his wife. One day in the kitchen, his wife remarked that she was frustrated because she had loaned a roaster oven to someone but couldn't remember to whom. Charles excused himself for a moment, disappeared into his study, and soon returned with the name of the friend who had borrowed the roaster—and the date.

Amazed, his wife asked how he could possibly know. Charles took her into his study and pointed to the bottom shelf of a bookcase that ran the length of the room. There, in a neat row of identical volumes, were Charles's accumulated diaries covering the years since his retirement. With the increased free time of retirement, Charles had decided to keep a diary. He began by noting details of life as minute and mundane as the roaster's migration. But he eventually expanded his notes to include more thoughtful observations and reflections on life. What had begun as a simple accounting—which for Charles was a familiar way to relate to the world—had evolved into a more textured account of his life, colorful, and with meticulous (and often helpful) detail.

Autobiographical writing and storytelling among older adults is common, and it stems from a variety of impulses, some psychological (such as a desire to find meaning and patterns in a long life) and some physical. Autobiographical expression in the second half of life appears to be an example of bilateral involvement of the brain hemispheres.

In one study, researchers examined brain activity among young adults and older adults as they recalled specific autobiographical events. The brain region of interest in the study was the memory-critical hippocampus. The two age groups showed comparable efficiency of recall, but the younger adults primarily used the left hippocampus, whereas the older adults used both the left and right hippocampi.

Perhaps part of the autobiographical drive among older people is related to a rearrangement of brain functions that makes it easier to merge the speech, language, and sequential thinking typical of the left hemisphere with the creative, synthesizing right hemisphere.

We have a lot to learn about how older adults might take advantage of research findings on the use of both brain hemispheres with age. Perhaps specific tasks or activities encourage hemisphere sharing. We don't know yet, but we do know quite a lot about ways to maintain and improve *overall* brain fitness, and these insights will undoubtedly support the kind of age-related bilateral brain use described above.

BRAIN FITNESS

We've seen that the older brain is more resilient, adaptable, and capable than we thought, and we've learned about four key brain attributes that support a more optimistic view of human potential in the second half of life: brain resculpting as a result of new experience and learning, new brain cell formation, maturing of the emotional circuitry, and bilateral activity in the aging brain. Now we can focus on how to take advantage of these brain processes, both to improve overall brain fitness and to protect our neural architecture from damage by disease or trauma.

Research has identified five categories of activity that, if practiced regularly, can significantly boost the power, clarity, and subtlety of the brain and mind. Engaging in any of them is for your brain what going to the gym is for your body—a healthy workout that releases and expands your mental potential.

Exercise Mentally

The brain is like a muscle. Use it, and it grows stronger, but let it idle, and it'll grow flabby. The findings we explored earlier about brain plasticity have spawned the field of behavioral neuro-science—the study of how external stimulation alters brain structure and functioning. One of the cornerstones of the field is the proven finding that experience modifies brain structure at every stage of life, from before birth to death.

[1]Behavioral neuroscientists, such as Joseph LeDoux at New York University, tell us that engaging in challenging new learning experiences boosts the development of the brain in the second half of life because the new experiences generate new synapses and other neural structures. This development improves information processing and memory storage, especially in the hippocampus.

You don't have to be an Estelle Jansen and enroll in graduate school at seventy to accomplish this. Many mentally challenging activities can stimulate your mind, such as community educational courses, book club discussion groups, writing groups, art programs, and challenging work—paid or volunteer, full-time or part-time.

Choose something appealing and challenging—something you'll have to work at. Just as with physical exercise, you want to work up a mental sweat. And don't be surprised if, once you start, you want to do more. One of the programs I co-chair is the Cre-ativity Discovery Corps, in which we identify unrecognized talented older people in the community. A ninety-three-year-old woman we recently interviewed advised us that she might find scheduling the next interview difficult because she was very busy applying for a Ph.D. program. Now *that's* a mental workout!

Exercise Physically

A growing body of evidence clearly shows that physical exercise boosts brain power. This is particularly true when the exercise is aerobic—continuous, rhythmic exercise that uses large muscle groups. The positive effects of aerobic exercise are undoubtedly due to factors such as increased blood flow to the brain, the production of endorphins, better filtration of waste products from the brain, and increased brain oxygen levels. Numerous studies support the positive effects of physical exercise. Among them:

- A Canadian study showing that physically active people have a lower risk of developing cognitive impairment, Alzheimer's disease, and dementia. The study also found that the more people exercise, the lower their risk of sustaining brain damage from stroke.
- A study of a group of older women showing that those who increased their physical activity by walking had less cognitive decline and dementia over the next six to eight years
- Studies of older adults that found slower declines in the density of brain tissue in the cerebral cortex among those whose cardiovascular fitness was high
- Studies suggesting that cardiovascular fitness increases the number of connections between brain cells in the frontal part of the brain, perhaps by increasing the networks of fine blood vessels in those regions
- Studies showing better learning and performance among animals that have had regular exercise. These studies demonstrate that exercise stimulates the production of important

neurochemicals that increase brain cell survival, neural plasticity, and the development of new neurons.

Pick Challenging Leisure Activities

The types of leisure activities you pursue affect your brain fitness. A study of the connection between leisure activities and the risk of dementia and cognitive decline found the following to be the most effective activities (listed in order of the impact they have):

- Dancing
- Playing board games
- Playing a musical instrument
- Doing crossword puzzles
- Reading

The reduction in risk was related to the frequency of engagement. For example, among older people who did crossword puzzles four days a week, the risk of dementia was 47 percent lower than among those who did them only once a week. Another study found that knitting, doing odd jobs, gardening, and traveling also reduce the risk of dementia.

In light of what we know about the factors that influence brain plasticity, these studies suggest that challenging activities build up a reserve of dendrites and synapses. Such studies offer further evidence that mental and physical exercise improve overall brain fitness.

Achieve Mastery

Research on aging has uncovered a key variable: sense of control. Older persons who pursue activities in which they experience a sense of control and mastery are healthier both physically and mentally than those who do not. Such activities might include learning to play a musical instrument, taking up embroidery, becoming computer literate, or learning a new language. The possibilities are unlimited. Notably, the positive influence of sense of control on health becomes more pronounced in the second half of life.

Developing mastery in one area can produce feelings of empowerment that extend to other areas of your life, increasing your comfort level with exploring new challenges—and again, challenges stimulate brain health.

One way that mastery and a sense of accomplishment can improve mental fitness is by boosting the immune system. Research on the mind/body connection shows that positive emotions boost levels of beneficial immune system cells. Two types of cells in particular respond to positive feelings: T cells, which are white blood cells that orchestrate immune defenses, and natural killer cells, which are large white blood cells that attack tumor cells and infected body cells.

Establish Strong Social Networks

Networks of friends and family and active social engagement are associated with better mental and physical health and lower death rates. For example, maintaining social relationships in the second half of life has been associated with reduced blood pressure, which in turn reduces the risk of stroke and the resulting brain damage. Maintaining social relationships also reduces stress and the harmful effects it has on the body. Stress hormones erode brain architecture, impair the immune system, burden the heart, and weaken resistance to mental disorders such as anxiety and depression.

Social activity also combats loneliness, which is more prevalent with aging. Loneliness, in turn, is associated with a range of adverse health effects, including slower recovery from coronary artery bypass surgery, more office visits to physicians, poorer dental health, and a greater likelihood of nursing home admission. Thus the positive impact of social networks on the health of the mind, body, and brain in late life can be profound.

In this chapter we've learned that healthy older brains are more robust and have more potential than most people realize. But this isn't the end of the good news. The brain is in some ways like the foundation of a building. It provides the physical substrate of the mind, the personality, and one's sense of self. Our brain "hardware" is capable of adapting, growing, and becoming more complex and integrated with age. But at the same time, our *minds* also grow and evolve. This is psychological development—development of insight, emotional stability, knowledge, creativity, and expressive abilities. Changes in the brain can strongly influence this development, for better or worse. And the reverse is also true: how we evolve and grow psychologically affects the brain itself, forging the same kinds of new connections and constellations of networks that new experiences of any kind create. Let's turn now to this relationship and the intriguing subject of psychological development in general.

2

Harnessing
Developmental Intelligence

All that is valuable in human society depends upon the
opportunity for development accorded the individual.
—Albert Einstein

CONTRARY TO POPULAR ASSUMPTIONS, development is not just for
kids. It doesn't end one step beyond adolescence, at the "destination" of adulthood. This is not nearly as obvious as it may sound.
As we saw in the introduction, the field of developmental psychology was largely based on theories that looked no further than the
onset of adulthood, as though all the important phases of growth
were finished by that time. And the popular view of aging is one of
loss, as though the developmental clock begins to run backward
from adulthood, eventually arriving back at childhood. Successful
aging in this view, means holding on to preexisting strengths and

capacities for as long as possible and slowing what is assumed to be an inexorable decline.

Of course, it's true that older adults in the grip of mental and physical illness can display a retrograde "development" in which they become more and more helpless and dependent, like children. Older adults in late-phase Alzheimer's disease, at its extreme, lose control of their bodies and minds and become, in effect, old babies in need of constant attention and support. It is also obviously true that from a purely physical standpoint, many body systems slowly weaken with age and lose some resilience and capacity.

But to view the second half of life through these narrow windows is to miss the most important aspects of aging and accept an unnecessarily dismal and depressing paradigm.

The larger truth about aging is that development, in the broadest sense, can be continuous, invigorating, and deeply rewarding. Why? Because the wellspring of growth and change never dry up. Development is the steady unfolding of an individual human being's full physical, mental, emotional, and philosophical potential. It is driven by a host of forces that wax and wane throughout life. Some of these drives are purely physical, such as the surges of hormones that propel both fetal and teen-year development. Some are emotional: at every stage of life, we crave love, acceptance, and attention from others and feel urges to love, accept, and pay attention to others. And as the most social species on the planet, we are strongly affected by our drives for social standing, for "success," and for expanding our sphere of control.

We share many of these drives with other animals, particularly other mammals. But human development is also powerfully fueled by drives originating in our capacities for abstract thought, reflection, creativity, and culture, which are, in turn, supported by our large, complex brains. We are an intensely curious species, for example. Curiosity about the world is one of those drives that can grow stronger with time, not weaker. The more we know, the more we realize how much we *don't* know. Curiosity fuels more curiosity, which, in turn, drives learning. As long as curiosity is not shut down prematurely by dogma, orthodoxy, or overly pat belief systems, it can flourish throughout life and be a fount of energy, vitality, and satisfaction.

People vary more in the "higher" drives of human development, such as spirituality and artistic expression, than in their more basic drives, such as those for comfort and security. Some people, for example, have a strong urge to be creative, whether in the classic arts or simply in whatever tasks or opportunities surround them. Others are not so consciously focused on creativity but feel a powerful drive to give of themselves to others. Some people are driven by a spiritual quest—a desire to find a spirituality that feels true to them and is compatible with their other beliefs about life. Of course, a person could feel each of these three drives equally powerfully. My point is simply that people can and do vary in their experience of the range of urges that drive human development.

In later chapters I explore some specific developmental drives. Here I want to lay down the foundational principle that our growth does not stop at "adulthood" but can continue throughout our lives. This growth is fueled by many strands of urge, desire, craving, longing, and seeking, which, collectively, I call the Inner Push. As I noted in the introduction, the Inner Push is a life force composed of many individual forces, like springtime sap rising through the myriad channels and pores inside a tree, propelling its flowering and seasonal growth.

In my work with thousands of adults from midlife to past age 100, I have seen these developmental imperatives emerge as most common:

- To finally get to know oneself and be comfortable with oneself
- To learn how to live well
- To have good judgment
- To feel whole—psychologically, interpersonally, spiritually—despite loss and pain
- To live life to the fullest right to the end
- To give to others, one's family, and community
- To tell one's story
- To continue the process of discovery and change
- To remain hopeful despite adversity

This is the developmental agenda for adult life, and I don't believe that Nature sets us up for failure. Whatever our challenges, the undercurrents of developmental energy support our growth toward these goals in different ways as we age.

I saw the Inner Push in action in the life of Kathleen Kramer, who was forty-one when I first met her. We were meeting to review the care needs of her seventy-three-year-old mother, who had Alzheimer's disease. In the course of talking with Kathleen, I learned something of her personal story. She had married hastily during her first year of college. She was pregnant and had her child within a year. Unfortunately, her husband was abusive, so, despite the burden of caring for a newborn, she decided to divorce.

The responsibilities of single parenthood forced Kathleen to leave college. She always loved to read, and she found part-time work at a bookstore. Reading to her young child ignited in her an interest in children's books, and she eventually became the resource person in the children's section of the bookstore.

As she approached forty, with her child in college and her mother finally being cared for in her own home, with outside help, Kathleen felt restless. She had always regretted dropping out of college but wasn't sure what she would study if she returned. Talking with me about her internal debate one day, she mentioned her love of children's books and reading. "Hmmm . . . " I mused, "are those random dots, or can you connect them into a larger picture?"

She chuckled and then said, "Well, perhaps I could go to college, become an English major, and have a special focus on children's literature."

I lost track of Kathleen after that because her mother was in good hands. But eleven years later Kathleen called me. Her eighty-six-year-old father was depressed. In the course of helping her map out a strategy for her father, I caught up on her life changes. She had acted on that earlier conversation, returning to college and getting a bachelor's degree and then a doctorate in children's literature. Kathleen started teaching children's literature at a local college and had just finished writing her first children's storybook.

Kathleen's story is increasingly common. A 2003 report from the National Center for Education Statistics shows that 16 percent of people obtaining a bachelor's degree were over age thirty. The

Institute for Higher Education Policy reports that the proportion of students age forty and older who enrolled in colleges increased by 235 percent between 1970 and 1993. Today, 10 percent of all undergraduates and 22 percent of all graduate students are forty and older.

But more fundamentally, Kathleen's story shows how our Inner Push to grow and develop never stops. She responded to the deep urgings of her psyche with energy and confidence, and her human potential blossomed.

DEVELOPMENTAL INTELLIGENCE

Development is not a race or a competition. No single goal exists that, when obtained, qualifies a person to say, "I'm now fully developed." It's true that most people pass through various stages and, in that sense, are "finished" with certain aspects of development. Learning to walk and talk, for example, are obvious stages that most people accomplish. As we age, the drives that fuel our development may be manifested to a greater or lesser extent, influenced by many factors, most of them not in our control. A person's innate need for attachment, for example, may be thwarted by a distant or dysfunctional parent. Our inborn curiosity about the world might be stunted by authority figures who discourage questioning or skepticism. Alternatively, our innate drives can be nourished and supported, which can produce a rich developmental flowering.

Development is highly individual, as Einstein's quote at the beginning of the chapter captures. The goal, ultimately, is simply to manifest your own unique potential. It is in this sense that we can use the term *developmental intelligence*—the degree to which a person has manifested his or her unique neurological, emotional, intellectual, and psychological capacities. To be developmentally intelligent means being aware of your own development, both past and present. It is also a description of your current developmental state—although this description isn't as precise as an IQ score.

Developmental intelligence is defined as the maturing of cognition, emotional intelligence, judgment, social skills, life experience, and consciousness and their integration and synergy. With aging, each of these individual components of developmental intelligence continues to mature, as does the process of integrating each with the others. This is why many older adults continue functioning at very high intellectual levels and display the age-dependent quality of wisdom (which I will discuss in chapter 6).

As I have emphasized from the start, there is no denying that problems can accompany aging—and research to date has focused mostly on such problems, typically in individual components of our total mental superstructure. Less attention has been paid to how gains and losses can occur at the same time. For example, older adults often experience more trouble with word finding—the "tip-of-the-tongue" experience—but at the same time, the total number of words they use—their vocabulary—continues to increase. If you look only at selected functions, such as certain aspects of memory or mathematical ability, you miss the larger picture of how functions become more integrated, often improving overall performance. This is the heart of developmental intelligence.

I am not as interested in *measuring* developmental intelligence as I am in conveying the idea that we all have, to one degree or another, developmental intelligence and, as with all intelligence, we can actively work to promote its growth. As we grow and evolve, developmental intelligence is expressed in the deepening qualities of wisdom, judgment, perspective, and vision. Some in my field use the term "postformal thought" to describe attributes that I associate with more advanced cognition, a component of developmental intelligence that becomes more apparent as we enter middle age. Another phrase that describes the same thing is "higher-level reasoning." This more advanced style of cognition is characterized by three related thinking styles:

- *Relativistic thinking* means understanding that knowledge sometimes reflects on our subjective perspective, that the context of a situation influences our conclusions, that contexts can change, and that answers are not absolute. Relativistic thinking involves synthesizing knowledge from disparate or opposing views. (This type of thinking is also sometimes called dialectical thinking.) Here's an example: *"Our relationship is complicated. I cycle quickly between loving him and wanting to leave him. But I've come to realize that I need to work on both of these feelings at the same time. By looking at why I love him, I can better understand what is good in the relationship and work with him to build on that. By looking at why I want to leave him, I can identify what's wrong and work with him on modifying that. I guess at forty-five I'm becoming older and wiser. It still may not work out, but I feel that I have a better perspective on things, and that will improve my odds of moving things in a better direction."*

- *Dualistic thinking* is the ability to uncover and resolve contradictions in opposing and seemingly incompatible views. It is the capacity to suspend judgment long enough to hold mutually exclusive views in mind at the same time. *"Trying something new has always been hard for me. I'm afraid of making the wrong decision, so I've sometimes waited too long and the opportunity passes me by. I have these dueling voices in my mind. One says, 'Look before you leap,' but in the past, I looked too long. The other says, 'He who hesitates is lost,' and there are times when I rush into things and hardly look at all. I used to think these two views were incompatible, but now I can take something from both—I think that together they are moving me off the dime."*

- *Systematic thinking* means being able to see the forest instead of the trees. It's an ability to pull back from an idea or situation to take a broader view of the entire system of knowledge, ideas, and context that are involved. *"I used to think that the lifelong conflict with my sister stemmed from the fundamentally different views we held. But then I came to realize that all our lives we'd been treated very differently by our parents, and I saw there were broader family dynamics involved. This gave me a new perspective on how I could approach her. It took me until my forties to fully see this. I realized that I had to look at our family system more broadly to understand how to relate to her better one-on-one."*

These three types of thinking are all "advanced" in the sense that they do not come naturally during our youth. By our nature, we prefer that the answers to our questions be black or white, right or wrong. We want a clear answer, and often any answer is better than no answer. We are notoriously uncomfortable with ambiguity, uncertainty, and contradiction. We constantly (and mostly unconsciously) try to avoid or suppress "cognitive dissonance"—the internal discomfort that arises from conflicting desires, ideas, or beliefs.

It takes time, experience, and learning to develop the capacities for relativistic, dualistic, and systematic thinking. It can be difficult to challenge existing beliefs that offer comforting, but dubious, answers to life's problems. It's sometimes hard to say, "I don't know" instead of, "The answer is . . . " But our capacity to accept some uncertainty, to admit that answers *are* often relative, and to suspend quick judgment for a more measured evaluation of opposing claims is a real measure of postformal thought and developmental intelligence. In fact, "wisdom" is in some ways a synonym for "developmental intelligence." Wisdom is how developmental intelligence reveals itself. Both are manifestations of mature thinking, accumulated life experiences, emotional development, and, as we saw in the previous chapter, positive changes in the brain that occur with aging.

An example of developmental intelligence in action comes from an editor I know in a New York City publishing company. He was in his sixties when he remarked to me one day over lunch that it hardly seemed right to start considering retirement when it had taken him forty years to finally grow up on the job.

This man had worked his way up through a string of positions over the years. His sharp intellect and passion for the work had been his greatest assets, but his impatience and lack of people skills had been a constant challenge. He had invaluable skills as an editor, but in job after job he struggled with interpersonal relationships because of his tendency to be brusque, critical, and insensitive to others' feelings. It had just been in the past couple of years, he said, that he was beginning to master interpersonal communication, a gift that some people seem born with but that can take many decades for others to master. As his emotional development caught up to his intellectual development, he morphed from a brilliant but brittle loner into a mentor and a mediator of interpersonal conflicts. "I feel like a changed man," he said, with a bemused smile. Age had worked its magic. The rich tapestry of neurological and psychological changes over the years had allowed the different strands of his developmental intelligence to mature and grow.

UNDERSTANDING OUR
DEVELOPMENTAL POTENTIAL

Development is as powerful in adulthood as it is in youth—with one critical difference. In youth we are largely powerless to steer our emotional and intellectual growth. As adults we have much more control. We can understand ourselves, our motives, and our challenges in ways a child cannot. Moreover, as adults, we have the personal freedom to make conscious choices and take purposeful steps to steer ourselves in a desired direction. Kathleen Kramer's story of earning a doctorate in children's literature is a fine illustration of that fact. Here's another example: If you know that your discomfort at making presentations is holding you back in a career you want to pursue, you can decide to take a public-speaking class or get some coaching. Similarly, if you feel your marriage fading, you can take steps to recharge the emotional connections with your spouse. In all kinds of circumstances, we can use our self-knowledge and our Inner Push to pursue positive changes. In a sense, we can hot-wire the brain and the mind to connect our reasoning and decision-making capabilities with our developmental energy, our internal power source—the Inner Push.

In the previous chapter we saw that age-related brain changes support continuing development in later life. These underlying physiological changes also support a host of psychological factors that hold the keys to positive change with aging.

Extensive research shows that long-standing psychological issues, such as anger, social anxiety, feelings of inferiority, and low self-esteem, are not immutable personality flaws. They can change, and in fact age can catalyze such change. A critical or hypersensitive sibling can shed that negativity and engage his or her brothers and sisters in more positive ways. The controlling parent can learn to love and let go. Shyness can be overcome, impulsivity can be reined in, and insensitivity can be softened.

Some traditional psychological theories stress the importance of "crisis resolution" in growth. Adherents assert that we must fully accomplish or resolve one developmental step in order to advance to the next one. New research, including my own studies of older adults, reveals a more complex picture. Phases (or stages) can ebb and flow, overlap, and follow one another in unpredictable ways as our Inner Push interacts with life events. Illness or other types of setbacks, for instance, can temporarily halt growth. We may need to "revert" to a previous behavior pattern, such as being dependent on others or needing to relearn certain skills. On the other hand, it doesn't appear necessary for all our psychological issues or problems to be resolved before we can enter and experience new phases. Some people live with long-standing relationship problems and nonetheless continue to flourish and grow in other areas of their lives, such as in creativity or social activities.

With a better understanding of adult development fueled by the natural impetus for growth from the Inner Push, we can develop strategies for creating more dynamic and satisfying lives. Adulthood is not a developmental destination or pinnacle, either psychologically or neurologically. It is the continuing evolution of our brains and our selves that constantly invites us to have a more active hand in designing our destiny.

As we saw in the introduction, many of the greatest thinkers in the field of psychology viewed growth and development as essentially complete by the end of adolescence or the early twenties. Two of the most influential scientific leaders—Sigmund Freud and Jean Piaget held this view. Even the great modern developmental psychologist Erik Erikson had only one stage after adulthood, which he referred to as "mature age." Moreover, none of his stages were linked to the new findings from brain research.

Erikson delineated eight stages of psychosocial development and defined each in terms of a psychosocial issue or conflict that must be resolved:

Erik Erikson's Stages of Psychosocial Issue or Crisis

1. Infancy	Trust vs. mistrust
2. Early childhood	Autonomy vs. shame, doubt
3. Play age	Initiative vs. guilt
4. School age	Industry vs. inferiority
5. Adolescence	Identity vs. identity diffusion
6. Young adult	Intimacy vs. isolation
7. Adulthood	Generativity vs. self-absorption
8. Mature age	Integrity vs. disgust, despair

Erikson and other classical thinkers believed that we go through each of these stages one at a time, one after another, and that the crises or challenges of each stage must be mastered before one can move to the next stage. (Freud included the resolution of certain psychosexual issues in his map of development.)

It's true that healthy development can be arrested by abuse, neglect, deprivation, or trauma, which can lock in an immature phase of psychological development. The failure to move through some of the very early phases, such as learning to trust, can interfere, or make more difficult, the more "advanced" later phases, such as developing a capacity for intimacy. But it's a mistake to view development as a straight-line phenomenon, in which failure to "succeed" at one developmental step necessarily blocks progress on other developmental tasks. It's more complex than that. The brain and the human spirit are more flexible and adaptable than some of these early theories suggest. I have seen many instances of people who have suffered horrendous early experiences and yet have their heads screwed on straight, are in healthy interpersonal relationships, and are doing fine things with their lives. This tells me that among the factors that shape our psychological development are built-in inner capacities that seek expression and can transcend adversity—the Inner Push I have been referring to.

In short, early theories of development are not necessarily wrong, they're just incomplete. This isn't an uncommon situation in science. Einstein's theories of relativity did not mean that Newton's earlier theories of motion and mechanics were wrong, just that they were limited and incomplete. Newton's laws work perfectly well for most terrestrial applications, but if you want to work at extreme scales of mass, distance, or energy, you have to use Einstein's equations. The same applies to human development. If you are interested in childhood or teen development, the theories of Erikson and other psychologists will work fine. But if we want to know about the whole human life span and, in particular, how older adults continue to grow and develop, we need a new, larger paradigm—one that does not require a lockstep view of human development.

The following is an example of a life situation that is more complicated than it would appear through the lens of Erikson's stages.

SALLY'S STORY

Sally DeMarco was approaching "the Big Four-O," as she called it, and came to me because all her romantic relationships had ended before marriage. She had grown up with a father whose job required frequent travel, often for weeks or months at a time. His schedule was unpredictable, so Sally never knew when he would be gone again. Now, with middle age looming, she was taking a hard look at her relationships, and she found some truth in something her friends had been telling her: the men she was attracted to were invariably distant—physically, emotionally, or both. For example, Sally's first major relationship had been with a correspondent for an international news agency. He was continually departing for trips abroad, was married to his work, and didn't want to settle down. Her next relationship was with a married man who said he was going to leave his wife. But it became clear he was more interested in working on a continuing set of home renovations than on breaking up his marriage. Later relationships had other insurmountable barriers to marriage—barriers that she was aware of at one level but that she chose to discount.

Now, nearing forty, Sally felt an inner drive to change her behavior. She recognized that her attraction to men who were hopelessly distant was related to her relationship to her own distant father. Viewed only through the psychosocial "crisis orientation" perspective, Sally's problems would be seen as a result of her failure to resolve the stage 6 psychosocial crisis—intimacy vs. isolation. Therapy would be aimed at helping her address these old wounds and transcend the emotional incompleteness they engendered.

Instead, I saw in Sally a preexisting motivation and will to break her old patterns, and this is what we focused on in therapy. She decided to try a strategy she had previously resisted—asking her friends to introduce her to eligible men, rather than randomly meeting them on her own. She also asked her friends to look for potential partners who did not seem to be in situations that would place hurdles in their paths right from the start. Two years later, she was engaged to Mike, a widower and successful businessman with no children.

Although it's true that to some extent Sally's choices in men were unconsciously influenced by her father's prolonged absences, that characterization is not broad enough to be truly helpful. Viewed through the lens of adult development as I've been describing it, Sally was acting on the positive Inner Push of a new developmental phase in her life, one that enabled her to stand back, draw on postformal thinking, and look at her situation differently. Doing so enabled her to make a different choice. Although she'd been adversely affected for years by her father's negative model and the associated trauma, she was able, through the positive dynamics of this emerging developmental phase, to make a midcourse correction in her life. She was determined to get out of a rut and move ahead positively.

In doing so, Sally exhibited an emerging developmental intelligence, which often does not fully kick in until early midlife. Psychotherapy can help people access their developmental intelligence, and this is what helped Sally. Like Sally, many of us have deeply rooted issues that require focused attention and the positive, creative urging of the Inner Push to resolve.

THE ROLE OF EMOTIONS IN DEVELOPMENTAL INTELLIGENCE

As we saw in the previous chapter, older adults experience less negative emotion, come to pay less attention to negative than to positive emotional stimuli, and are less likely to remember negative than positive emotional experiences. This maturation is related to changes in the brain's key emotion center, the amygdalae. Our neurological development lays the foundation for better control of anger with aging and improved conflict resolution capabilities. When dealing with conflict, adolescents and younger adults tend to use more outwardly aggressive and psychologically immature strategies, a tendency that reflects their generally lower levels of impulse control and self-awareness. Older adults are able to modulate their emotional responses and react with more care and greater awareness of both themselves and others.

Better emotional control can also result from improved self-awareness. For example, you may come to realize that competitive work environments aggravate your tendency toward depression. This awareness can allow you to choose different kinds of work environments. Similarly, knowing that large, impersonal parties make you feel uncomfortable and invisible allows you to screen out such events and choose ones more to your liking.

The improvement of this type of emotional intelligence (an important component of developmental intelligence) with age can be absolutely critical to achieving the peace of mind, sense of completion, and self-fulfillment that are the sweetest fruits of aging, as the following story illustrates.

ELLIS DANIELS: CHOOSING A LEGACY OF LOVE

Ellis Daniels had always been authoritarian, rigid, and short-fused. He displayed a combative, take-no-prisoners attitude with anyone who disagreed with him. This emotional volatility played a key role in one of his deepest wounds: the thirty-year estrangement from his son David, now fifty-one years old.

Ellis was a lawyer who had followed in the footsteps of his lawyer father and grandfather, and he had always assumed that David would carry on the family tradition. But David resisted, setting up confrontations and lingering resentments on both sides. He was artistic and entrepreneurial, and he dreamed of opening a restaurant and serving creative cuisine. Ellis was appalled, calling David's aspirations "flighty" and "ridiculous." Under intense pressure from Ellis, David applied to law school and was accepted. Ellis, thrilled, planned a gala event to celebrate, and in invitations to family and friends he proudly boasted of his fourth-generation future barrister.

But David didn't show up for the party in his honor. Later, confronted by his father, David told him that he'd decided against attending law school in favor of pursuing his dream. Ellis flew into a rage and as much as disowned him. David, offended by the threats and intimidation, left home and tapped his savings to open a restaurant.

Ten years later, David's restaurant was thriving, and he opened a second, and then a third. Despite his success, Ellis held a grudge and refused to move toward reconciliation. When David and his wife, Ellen, had children, they made sure their children spent time with their grandfather, but David was never part of the visits.

It was during preparations for Ellis's eightieth birthday that he began to question his decades-long hostility. Ellis's daughter, Darlene, made her position clear: "You've always been such a responsible man, a great lawyer, and a terrific father," she said. "Do you really want your legacy to include a thirty-year feud with your only son?" She suggested that he could honor the family patriarchs in other ways, by telling their stories in a memoir, for instance, perhaps inspiring future lawyers in the family.

Ellis admitted that over time he'd come to feel less angry and bitter about David's choices. A short time later, when David invited the family to a graduation celebration for one of his daughters, Ellis made a new choice. He showed up, and the handshake he extended to his son turned into a father-son hug and the start of a new era in their relationship and for the family.

Ellis was responding to the Inner Push to review, give back, and confront unresolved conflicts. He had grown more developmentally intelligent over the years, and that included being less emotionally reactive and more open to changing his mind about his son's choices. A seemingly simple reconciliation like Ellis's requires layer upon layer of change, a reorientation based on shifts in his personal values, emotional maturity, and conflict resolution skills.

THE FOUR HUMAN POTENTIAL PHASES IN THE SECOND HALF OF LIFE

We've seen that we are all endowed with a host of internal drives, which I call the Inner Push, that fuel our development and can provide the energy we need for making positive changes. How we experience this Inner Push and how we express it changes as we age and is influenced by ongoing brain development as well as our accumulated life experience. In my research, I've discovered some broad regularities and patterns in the resulting psychological growth and development of older adults, which I have articulated as four developmental phases of the second half of life. The phases are commonly but not always separated by time; sometimes they coexist, intersect, and interact with one another. Each phase holds surprising potential for positive growth, as we will see in the next two chapters.

3

The Second Half of Life:
Phases I and II

The best way to predict the future is to create it.
—Peter Drucker

PSYCHOLOGY, AS WE'VE SEEN, has severely underestimated the positive potential of the second half of life. We now know that the human brain can learn at any age and that it remodels itself in ways that can make an older brain more powerful than a younger brain. Our psychological, social, intellectual, and emotional development never stops, either. Indeed, the forces I describe as the Inner Push fuel continued developmental intelligence through a range of important life transitions.

After talking with thousands of older adults, I have formulated a new—and empowering—view of the second half of life. Picking up where Erikson left off, I have divided his expansive "mature age" stage into four phases of growth and development. This chapter explores the first two phases, but first let me give you the big-picture perspective by outlining the four phases.

Each phase is named by three formal descriptors as well as a shorthand title, and each occurs during a characteristic age span.

Phase I: Reevaluation, exploration, and transition (*midlife reevaluation*); mid-thirties to mid-sixties but usually occurring during one's early forties to late fifties.

- People in this phase seriously confront their sense of mortality for the first time.
- Plans and actions are shaped by a sense quest or, less commonly, crisis.
- The brain changes during this phase spur developmental intelligence, which is the basis for wisdom.

Phase II: Liberation, experimentation, and innovation (*liberation*); mid-fifties to mid-seventies but usually occurring during one's late fifties to early seventies.

- People in this phase often have an "if not now, when?" feeling that fosters the new sense of inner liberation.
- Plans and actions are shaped by a new sense of personal freedom to speak one's mind and act according to one's needs.
- New neuron formation in the information processing part of the brain is associated with a desire for novelty.
- Retirement or partial retirement gives people time to experiment with new experiences.

Phase III: Recapitulation, resolution, and contribution *(summing up)*; late sixties into the nineties but usually occurring during the late sixties through the eighties.

- People are motivated to share their wisdom.
- Plans and actions are shaped by the desire to find meaning in life as we look back, reexamine, and sum up.
- Bilateral involvement of the hippocampi contribute to our capacity for autobiographical expression.
- People in this phase often feel compelled to attend to unfinished business or unresolved conflicts.

Phase IV: Continuation, reflection, and celebration *(encore)*; late seventies to end of life.

- Plans and actions are shaped by the desire to restate and reaffirm major themes in our lives but also to explore novel variations on those themes.
- Further changes in the amygdalae promote positive emotions and morale.
- The desire to live well to the very end has a positive impact on family and community.

As you can see, these phases can overlap significantly, and people can pass through them at different speeds. Some people, for instance, are confronted with their mortality in youth, whether by the premature death of parents or other loved ones or by near-death accidents or illnesses. Such people may experience some or all of the attributes of a life reevaluation that more typically occurs in older adults. Real life—and real people—are always more complicated than the neat structures we can create in books. But talking about these phases as separate entities is helpful because it allows us to see them more clearly and understand their nuances and attributes. In a sense, my four phases are like any other model in science: a simplification that allows us to easily grasp a truth about reality. DNA, for example, is not the tidy spiral depicted in textbooks—it's quite messy in fact, and the strands loop, kink, split, and interact in a bewildering manner. But the simple model allows us to better understand this all-important foundation of our beings.

When I present my vision of the four phases at conferences, people tell me that they see these phases mirrored in their own lives or the lives of family members or friends, and that the theory helps illuminate the various developmental strands of the second half of life. As you read about the phases, you may feel the same way, or you may find that your own experience doesn't quite align with what I describe. That's to be expected. We are each unique and no single framework will fit everyone. But these development strands are universal. They have the potential to make the hallmark of older adulthood not a time of stasis or decline but a dynamic time of growth, learning, and deep satisfactions.

My Own Transitions

I was taken by surprise the first time a colleague gently asked if I might be going through some sort of midlife crisis. I was puzzled because I didn't feel at all "in crisis." Rather, I felt newly challenged, invigorated, and in touch with new aspects of myself. Over the next couple of years, several more colleagues, some awkwardly, others with humor, made the same diagnosis: midlife crisis. They were all responding to my newfound interest in designing games for older adults.

One of the more clever ways the question was posed was by a friend who jested, "Are you turning *right* on us?" Now, in Washington, D.C., "right" invariably means "Republican." But my friend is a neuroscientist, and he was actually referring to the right side of the brain. He was suggesting that perhaps I had traded my logical, analytic abilities for the supposedly looser, more creative, and less disciplined abilities typically viewed as the province of the right side of the brain.

The truth, of course, was not nearly as simple as my friend's comment suggested. I was in my late forties, and I *had* become a game inventor, but I hadn't abandoned my career in gerontology. And I wasn't trading one side of my brain for the other. Rather, I was responding to my own Inner Push by creating opportunities for myself that took advantage of both sides of my brain and my own budding developmental intelligence. My game designs, initiated during my last three years of a twenty-year career at the National Institutes of Health (NIH), drew on lessons from gerontology and also promoted intergenerational play. During that same period I had become president of the Gerontological Society of America and founded an innovative research center at George Washington University—the Center on Aging, Health & Humanities. But my game inventions generated publicity, which, to my friends, suggested an entirely new persona or a response to a "midlife crisis."

My venture into game design was fueled in part by a personal confrontation with my own mortality. Two years before I left NIH I had been diagnosed with amyotrophic lateral sclerosis (ALS), more popularly known as Lou Gehrig's disease. ALS is a progressive and as-yet-incurable disease that slowly steals a person's control over his or her body. It was a very bleak diagnosis. I didn't sleep well at night and couldn't concentrate during the day. But I wasn't depressed—just filled with a dark restlessness. I had always fantasized about making a board game, and with the sentence of a life-degrading illness hanging over my head, I figured it was now or never.

I teamed up with a talented artist colleague, Gretchen Raber, and my first game was a finalist in an internationally juried show on games as works of art. I was hard at work on designing my second game as well as pursuing my career in gerontology when my doctor called with stunning news: I had been misdiagnosed. My symptoms were *not* caused by ALS. I had been visited by a sort of Dickensian archangel that prompted me to pursue a new path and awaken creative sides of myself that I had previously neglected.

These experiences left me pondering the similarities between what I had been through and the life reevaluations I was hearing from many of the people in my research study. I began to think about other life transitions I had witnessed among my many patients, research subjects, friends, and acquaintances and how those transitions were often mocked, pigeonholed, or trivialized. My transition had been labeled as a "crisis" by my own friends, even though it was nothing of the sort; the label seemed instead to make a cartoon out of a deeply meaningful and important phase of my life.

The more I thought about these ideas, the more I realized that our understanding of human development in the second half of life was woefully lacking. That was when I began formulating the ideas that resulted in this book. I wanted to help others who were going through the phases of later life to be more comfortable and less fearful and to seize opportunities when they presented themselves. I wanted to help us move beyond the distortions and myths that undermine individual initiative toward an understanding of the physiological and psychological mechanisms that enable us to design our own destiny. Over and over in my work with adults of all ages, I have seen that understanding these mechanisms and appreciating their reality can propel people toward positive growth.

PHASE I: MIDLIFE REEVALUATION

Middle age, for most of us, is the first time that we seriously consider our own mortality. We start for the first time to think about how many years we have left as opposed to how many have gone by. The fact of our own death strikes deeper than the abstract idea that everybody, of course, dies sometime. This profound new awareness fosters those classic questions: "Who am I?" "Where have I been?" "Where am I going?"

How deeply this recognition affects us depends critically on our beliefs about death and the degree to which we have already come to terms with it. I have found that even intelligent, self-aware people who can speak easily of death and do not fear it nonetheless reach a point in their lives where the prospect of their own demise acquires an emotional power that drives much deeper than their previous intellectual understanding. For others, the prospect of death is truly frightening—so frightening, in fact, that they avoid thinking about it much at all. Some people are comforted by a belief in an afterlife, reincarnation, or some other kind of life after death. Indeed, the diversity of beliefs about death and the range of ways human beings minimize or deny the reality of death are a powerful statement about the importance of this factor in our lives.

Research on aging has found that confronting one's mortality, along with other physical and emotional transitions of middle age, can provoke angst or profound anxiety. You can find echoes of this existential awareness in art, music, and literature from all ages. "Between my head and my hand, there is always the face of death," wrote Francis Picabia (1879–1953), the French painter and poet, when he was forty-four.

Unease about one's eventual death can sometimes produce physical symptoms such as restlessness, edginess, apprehension, irritability, muscle tension, fatigue, sleep problems, and depression. Unremitting angst raises the risk of developing clinical anxiety or depression. But normally the awareness that our life span is finite, regardless of what we believe will happen after death, generates a powerful strand in the Inner Push. Looking death square in the eye can spur a reevaluation of our priorities and goals. It can also foster a broader perspective on life that makes daily annoyances, accidents, and irritations less emotionally powerful.

It's no accident that one of the fundamental exercises in Zen Buddhism is something called a "death meditation." In nine specific points, it forces meditators to focus on their own death and the fact that the time of death is unpredictable. The meditation also asks meditators to ponder what things will help them when death does arrive. Although this exercise may strike Westerners as unusual, it is viewed in Zen as a natural and effective way to focus attention on the preciousness of life and on the aspects of life that provide true happiness and meaning.

The awareness of death, of course, is not the only salient feature of this phase of life. Reevaluation can be triggered by other life changes: the departure of children from the home, the experience of an accident, illness, or (as my example showed) a diagnosis of a serious medical problem. The process of reevaluation often feels like the beginning of a journey, which, of course, is exactly what it can be.

ALEX HALEY'S STORY

Alex Haley was approaching forty and was feeling grumpy and restless—itchy for something, but he didn't know what. He had enlisted in the Coast Guard at seventeen and stayed there, working in the public relations department as a writer. But he didn't want to write press releases all his life. After twenty years in the Guard, spurred by an Inner Push to do more, he quit.

"I wasn't going to be one of those people who died wondering what if . . . ?" he said in a published interview. "I kept putting my dreams to the test—even though it meant living with uncertainty and fear of failure."

He pursued his freelance career vigorously, writing for *Reader's Digest* and *Playboy*, and with the publication of *The Autobiography of Malcolm X*, which he cowrote, his career was taking off. Still, he felt a drive for something else.

In 1965, at the age of forty-four, Haley stumbled upon the names of his maternal great-grandparents when he was going through post–Civil War records in the National Archives in Washington, D.C. This began an eleven-year odyssey of research and writing about his ancestry, which culminated in the publication of *Roots* when Haley was fifty-five. The book was a best seller and was then turned into a television miniseries that stirred a nationwide interest in black history.

"In all of us there is hunger, marrow-deep, to know our heritage—to know who we are and where we come from," he said. "Without this enriching knowledge, there is a hollow yearning. No matter what our attainments in life, there is still a vacuum, emptiness, and the most disquieting loneliness."

The Myth of the Midlife Crisis

As I noted earlier, it's a misconception to think that everyone goes through a "midlife crisis." I have found instead that most adults experience a sense of quest in midlife. The insightful reflections that are common in this phase can provoke a powerful desire to find meaning in life, to begin new works, or to take existing works in new directions. Both the decade-long Successful Midlife Development study by the MacArthur Foundation Research Network and my own ongoing research on retirement-age individuals show that far fewer people experience what they call a "midlife crisis" than had been previously reported. Rather than leading to crisis, midlife reevaluation more characteristically leads to a sense of personal discovery.

"I'm looking forward to pursuing the career I always wanted," one forty-nine-year-old woman told me. "I'm tired of just working on other people's visions, rather than my own, even if I have to start on a smaller scale." A man in his early fifties who had gotten involved in his first major volunteer project said, "You know, I used to think the only way I could make a difference was through my work. But community involvement can be just as powerful a path."

One reason that the notion of a "midlife crisis" arose was that in the days before psychotherapy was so widely used, it was often only at middle age that men, in particular, would seek counseling for emotional distress, typically after years of emotional struggle or failed relationships. It wasn't necessarily that middle age produced the crisis, but that the emerging middle-aged mind told them to *do* something about it. And when they did, their efforts were misconstrued as a crisis.

The contrast between the view of midlife as a quest and as a crisis can be seen by looking at how each affects some key aspects of life. In a crisis, for example, our thoughts are disturbed and concentration is difficult. In a quest, however, our thoughts can be highly focused as we explore the nature of a problem or a possible solution. Emotions follow a similar pattern. In a crisis, depression is common and emotions tend to be volatile. In a quest, emotions may swing, but not out of control, and one retains the incentive and drive to resolve any difficulties encountered.

Sometimes, of course, life presents you with a truly serious problem. Divorce, serious illness, the death of a spouse or cherished loved one, or an unexpected job loss can all throw you into crisis mode. At such times, seeking professional counseling and the advice of family and friends can help you control your thoughts, emotions, and behavior. You can slowly restore your equilibrium, get back into quest mode, and tap your developmental intelligence to rediscover your path in life. Such crises strike at any age, but midlife is not necessarily any more full of them than other ages. The mistaken view is not that crises do not occur in midlife, but that crises should be *expected* in midlife, or that a crisis is *necessary* for passage through this phase of life.

POSITIVE QUALITIES OF MIDLIFE

Research shows that midlife reevaluation usually results from a series of small steps and gradual awakenings that can produce the following positive qualities:

Less impulsive responses to situations and people in daily life. "In my twenties, no sooner would something be in my lung, then it would on my tongue," said a forty-eight-year-old woman who had difficulty censoring insensitive comments that hurt people's feelings. In her forties she learned to pause and collect her thoughts before blurting out something she might regret.

A more thoughtful perspective on work. "In my thirties, what mattered most was my salary and the possibility for promotion," said a fifty-five-year-old man. "Now, it's the quality of the work environment and the meaningfulness of my work that are most important."

Openness to new ideas or complexity in life. "I used to think there was only one right answer to every problem, but I've come to realize that it's not that simple," said a forty-two-year-old patient.

"Sometimes there's *no* right answer. Realizing this has made me a more tolerant person."

Greater respect for intuitive feelings. A fifty-one-year-old patient commented that in his twenties he frequently obsessed about decisions, looking for the most logical choice. But as he moved through his forties into his fifties, he found himself starting to be influenced as much by gut feelings as by logic. "It's funny," he said during one session, "on the one hand I feel some of my decisions are less objective and more subjective, but, on the other hand, I'm more comfortable with many of the choices that follow."

Brain Changes at Work in Midlife Reevaluation

As we saw in chapter 1, the brain remains capable of growth, maturation, and resilience throughout life, in the absence of disease or accident. Some of these changes are particularly relevant during the midlife reevaluation phase of life. For example, the increasing use of both sides of the brain for cognitive processes—bilateral brain involvement—can support a more balanced perspective on life that draws on both our logical, analytical powers as well as our nonverbal, intuitive capacities. This likely contributes to the "post-formal thinking" we talked about in the previous chapter, which is a foundation for developmental intelligence.

When you hear someone saying "My head tells me do this, but my heart tells me to do that," they are more likely to be in their twenties than their fifties because, with age, our heart and mind, thinking and feeling are usually more integrated with each other.

Evidence for this kind of development comes from studies such as that from the Berkeley Institute of Personality and Social Research of women in their forties and fifties. Compared with younger women, the midlife women in this study had a stronger sense of personal identity, better self-awareness in social environments, more confidence, more control over events in their lives, and greater productivity.

I want to close this discussion about the midlife reevaluation phase of life with two stories from my experience. Both illustrate how the Inner Push advances developmental intelligence and the midlife reevaluation that is the defining characteristic of this phase.

James Dunton at Midlife

When I first met James Dunton he was having frequent night-mares and was feeling a pervading sense of anxiety. A computer engineer and consultant to dot-com start-up companies, he was financially successful but socially frustrated. He worked constantly and traveled frequently. Although his drive propelled him to busi-ness success, it left little room for long-term relationships.

Just after he turned forty, he began to reexamine his life's path. At this time he also had a series of disturbing dreams in which elec-trical fires and other calamities destroyed computer-filled offices. He wondered if his dreams were telling him something. "Is this what I want to do for the rest of my life, be dominated by work?" he asked me. "Will I just keep doing this until I burn out?"

In therapy I helped him explore his feelings and encouraged his introspection. His dreams provided a valuable focus for his attention and, because of their emotional power, helped motivate the work.

Later that year, James received a call from a professor with whom he had studied at business school. The school was recruiting for a faculty position, and the professor was searching for someone with James's experience. "You've made enough money," the teacher said. "Why not have a change of pace and try being a teacher? You'd be great!"

At first James declined, but his nightmares returned with a vengeance. He decided to apply, despite the pay cut it would entail. After negotiating summers off, he took the job. Not only did he find the work satisfying, fifteen months later he was engaged to an assistant professor in the American Studies Program. Interestingly, her doctoral thesis was titled: "The Half Life of a Dot Commer."

James's trajectory of early success and midlife reevaluation is very common. The drives of his youth for money and status were slowly displaced by different drives—in his case, for a stable, loving relationship and a more balanced life. As his values shifted and the foundations of his previous assumptions crumbled, he entered a period of uncertainty and anxiety. This, too, is normal. The anxiety and ambivalence were part of his developmental process. He wisely sought help during this time, and, as a result, he was able to harness and understand his feelings rather than be dragged down by them.

Fulvia Ramirez at Midlife

Fulvia Ramirez, a participant in one of my studies, was a sophomore in high school when her father was killed in Vietnam. The oldest of six children, Fulvia suddenly had to help support the family. She gave up her dream of college, although she finished high school, working part-time as she did so.

When she graduated, she joined a company that valued her bilingualism in English and Spanish. She was well paid and was praised for her attitude and skills. At twenty-three, she married one of the managers at the company. She worked for another year, but then quit with the birth of her first child. She had two more children in the ensuing years, and as they entered school she worked part-time as a Spanish translator. But as her children grew up, Fulvia began to pay attention to some long-buried aspirations. She took continuing education courses and found that she was attracted to the field of psychology.

When her youngest daughter was looking at colleges, Fulvia read the college catalogs for herself. Two years later, just shy of her fiftieth birthday, Fulvia was accepted in a psychology program at a local college.

Fulvia's midlife reevaluation is a picture of maturing developmental intelligence. Her cognitive abilities, emotional intelligence, judgment, and social skills all became increasingly strong and integrated. I believe that Fulvia's sense of empowerment and conviction about going to college was fueled not only by the stirrings of the midlife reevaluation phase but also by the promptings of a new developmental phase—liberation—which is characterized by a strong sense of freedom to act on one's convictions.

PHASE II: LIBERATION

Jeannette Palmer had spent most of her working life ringing up sales in a department store. At sixty-six, when I interviewed her, she was effervescent and not in any mood to retire. "I'm itching to do something different," she said. "It needs to be something out of the ordinary—maybe even a little risky. I'm done working the register."

In the weeks that followed, Jeannette brainstormed possibilities. She wanted to use her existing strengths and do something she liked. Something with some pizzazz. Then it came to her: she liked to drive, and she liked talking with people.

"There are a lot of ladies older than me who are giving up driving, but they still want to get out and about," she told me. "Maybe they need a driver."

She ran a classified ad in the local paper: "Seasoned mid-sixties female driver available to chauffeur and provide interesting conversation for mature older women." She immediately received calls, and her new business was launched. She decided that, for security, she ought to carry a handgun. So she took a class, got a permit, and bought a pistol. She also found, during her drive-time conversations, that several women were, like her, avid poker players. So she organized a weekly poker night, which, as one of her friends put it, "Sure beats bingo!"

Jeannette, the pistol-packing, poker-playing lady chauffeur, kept her business going until she was in her mid-seventies, and said she never, for a second, regretted her "career move." I saw in her yearning to do "something different" the signs that she had reevaluated her life path and entered the liberation phase, as illustrated by her willingness to risk starting her own business. In this phase, patterns of thought and behavior are characterized by a newfound comfort with who we are and the courage to express ourselves freely or to try new things. Creative endeavors are charged with the added energy of personal freedom, both psychological and, because of retirement, literal. People in this phase often feel their previous inhibitions lift and are increasingly apt to ignore social conventions. The prevailing attitudes of the liberation phase are: "If not now, when?" "Why not?" and "What can they do to me?"

People tend to feel comfortable about themselves by this time, knowing that if they make a mistake it won't do any serious harm to their self-image. This provides a favorable context for experimentation and innovation. As one man put it, "You don't have to prove anything to anyone anymore. That's an emancipation." So, whereas someone in his twenties may not take an art class for fear of looking incompetent, the same person at midlife or older—especially as this liberation push grows in intensity—will be much less concerned about appearances and much more open to new learning. In this frame of mind, we're more likely to take a chance on a class, or a cruise, or some other novel experience. The power and comfort of these feelings is captured in Mark Twain's seventieth birthday essay:

> The seventieth birthday! It is the time of life when you arrive at a new and awful dignity; when you may throw aside the decent reserves which have oppressed you for a generation and stand unafraid and unabashed upon your seven-terraced summit and look down and teach—unrebuked.

Retirement

Retirement is the most widely shared rite of passage in American adult life for those in their mid-sixties and beyond. We'll explore new findings about retirement more fully in chapter 7, but first let's set the record straight: The developmental impetus for adults at or approaching retirement age is *not* toward retirement, disengagement, or dimming vigor. On the contrary, the universal Inner Push at this age is the emerging sense of personal liberation. And while retirement can catalyze this phase, it is not required to stir up feelings of freedom and exploration.

The liberation of retirement can resemble adolescence. Both periods are marked by a strong orientation toward experimentation with new roles and a new sense of autonomy. Both also represent significant shifts in personal identity. In adolescence, one's sense of self is powerfully shaped by hormones, education, and experience. In the adult liberation phase, new facets of personal identity can reveal themselves, or they can develop from scratch.

An Inner Push to test limits is particularly strong during both periods. In adulthood, this can express itself as a kind of devil-may-care, lighthearted behavior sometimes laughingly referred to as a "second childhood."

At a deeper level, this developmental energy combines with the specific patterns of brain growth we learned about in chapter 1, which strengthen the parts of the brain responsible for information processing, learning, and memory formation. In particular, dendrites, the branchlike extensions of neurons that facilitate communication between brain cells, reach their greatest number and density in the hippocampi of human beings from the early fifties to the late seventies, a period that completely encompasses the liberation phase. Moreover, new neurons continue to grow in the hippocampi. This combination of brainpower and psychological development provides the energy for exploring new challenges, learning new skills, and experimenting with new activities, roles, and relationships.

Annette Green: Older, Wiser, and Braver

Annette Green was never a high-powered individual. A kind, gentle person, she worked as a housekeeper and tended to defer to others and not "rock the boat." When she was sixty-eight, she came to me for advice about her daughter, who had a long history of mental illness. She told me she had always felt that her daughter's doctors knew best and that she should accept their advice and decisions. But recently she had begun feeling that she should be more assertive and be a stronger advocate for her daughter's care.

These were new feelings, she said. In fact, her visit with me was the first time she had sought a second opinion about anything regarding her daughter. But she seemed comfortable, sitting with me, asking for advice. During our conversation, she also mentioned a grandson who was succeeding despite the dismal environment of a beleaguered inner-city high school. Annette said she had always felt intimidated about exploring new opportunities for him, but lately she just couldn't sit by and not try to help. "If I don't do it, who will?" she said.

She left that day with contacts and recommendations from me for some community programs that I thought could buttress the care her daughter was already getting. I learned later that Annette successfully followed up on those suggestions and was pleased with the progress her daughter was making. I also learned that she had pursued some leads for private high schools and scholarships for which her grandson might be eligible. In the end, he was accepted on scholarship to one of those schools. She told me that she felt "on top of the world" about his accomplishment—and about her own newfound courage and success in acting on it.

If Annette had come to me years earlier, when she was not yet in the liberation phase, I doubt she would have felt the necessary freedom to be so proactive on behalf of her daughter and grandson. But the internal changes accompanying this phase created fertile ground for my suggestions. Her own Inner Push drove her to do the rest.

No Time Like the Present

"If not now, when?" is an oft-repeated mantra of the liberation phase. By this stage of life we have generally resolved any angst arising from our confrontation with mortality, and we feel an increasing need to use well the time that we have left. This perspective can powerfully spur doing something you always wanted to do but never had the time or courage to tackle.

Phil Smith was a successful corporate communications executive. Approaching sixty, he explained that he felt drawn to devoting more time to his two favorite activities: bicycling and photography. He decided to retire from full-time work and shift to consulting. As he devoted more time to photography, he gained a local reputation for portraying the rural landscapes in his area. When asked what prompted him to follow his passion for photography, he said he felt liberated from the concerns of youth by an awareness of the preciousness and finiteness of life.

"You also start to realize more fully that it's not the long sweeps of history that we're dealing with here," he said. "It's the individual days, and every one of them counts. Our job—no matter what our work may be—is to make those days as good as they can be, one by one by one."

The courage to pursue long-deferred dreams can also be seen in Emily Hale's story.

Emily had been a career librarian, a job that fit her soft-spoken personality and preferences for solitude and quiet. Emily also had an activist streak, and she had always supported the causes of public education, the environment, and conservation—though mostly by donating money, not speaking up at meetings or taking leadership roles.

When she retired at sixty-five, Emily told me she felt a growing voice inside pushing her to put her knowledge to use. She spent the next couple of years becoming familiar with local and state politics and working and writing more visibly in support of public education. Her work was noticed, and she accepted an appointment to a task force of her local board of education. She developed a particular passion for appropriate student/teacher ratios. She argued that the current system legally allowed too many children to be packed into a single classroom. In pushing for policy change on this issue, she said she surprised herself by her willingness to confront those who trivialized the issue or suggested that change was not feasible.

For the first time in her life, she found herself in heated discussions. One day, she told me she wondered whether her behavior would jeopardize her relationships with her colleagues or even her position on the task force. But then, answering her fears and articulating her growing inner voice, her tone turned steely. Tapping the table for emphasis, she said, "Well, but it's the right thing to do, and besides, I'm sixty-eight years old—what can they do to me?" Then she smiled.

REEXAMINATION AND LIBERATION

In this chapter we've explored the first two phases of older age. We've seen how a new awareness of our own death can propel us into the phase of reexamination and how both psychological and physical changes in our minds and brains can free us with a new sense of liberation. These two phases of life can be richly productive. I've described ordinary people whose lives today illustrate this truth, but it's easy to find similar stories throughout history. Socrates was seventy when he was forced to commit suicide because his ideas threatened the assumptions and beliefs of the rulers of Athens. It was also at age seventy that Nicholaus Copernicus published his evidence that the Earth revolves around the sun, which provoked revolutions in both science and theology. Later, Galileo, at sixty-eight, extended and forcibly argued for the truth of Copernicus's theory. He was promptly arrested and spent the last eight years of his life under house arrest. To pick just one of countless stories that could be pulled from more recent times, Laura Ingalls Wilder started writing her *Little House on the Prairie* series of books at sixty-five. She continued writing for the next ten years.

I want to stress again that the midlife reexamination and liberation phases are not "crises," nor are they the result of crises. Of course, the growth and change that occur during these phases often involve some anxiety, discomfort, hesitancy, or confusion. That's normal, but these phases are, on the whole, phenomenally *positive* experiences, unless one is suffering from disease or extreme loss or privation. The latter two phases of life have an equal potential for positive growth, as we'll see in the next chapter.

4

The Second Half of Life:
Phases III and IV

In this world it is not what we take up,
but what we give up, that makes us rich.
—Henry Ward Beecher

PHASE III: SUMMING UP

The third of the four phases of later life typically occurs as one approaches seventy, although it can be present in different forms a decade earlier or later. It usually follows the midlife reevaluation and subsequent sense of liberation that I discussed in the previous chapter. I call this third phase "summing up" because people in this phase feel more urgently the desire to find larger meaning in the story of their lives through a process of review, summarizing, and giving back. In the summing up phase, we begin to experience

ourselves as "keepers of the culture" and often want to contribute
to others more of whatever wisdom and wealth we may have accumulated. I have seen countless people in this phase act on a
growing desire to give back through volunteerism, community
activism, and philanthropy. The Inner Push to sum up is often
expressed creatively through a recapitulation and review of one's
life through personal storytelling, memoirs, and autobiography. In
the retirement study I am conducting, the majority of participants
in their seventies and eighties are writing memoirs, doing oral histories, compiling photographs, creating family genealogies, or
producing other types of life review. A high-profile example of this
expression is the memoir written by Katharine Graham, the former
publisher of the *Washington Post*. It was her first book, and she
wrote it at age seventy-nine. (It won a Pulitzer Prize in 1998.)

I believe that the autobiographical urge in this phase is fueled
both by a continuing awareness of one's mortality and by some of
the physiological changes in the brain that were discussed in chapter 1. In particular, it seems to me that the use of both sides of the
brain allows for an optimum expression of the full range of factual
and emotional elements in a person's life story. I don't think it's
completely accidental that older people enjoy telling tales from
their lives so much or that life review is so popular at this age. It is
undoubtedly true that having more free time plays a role here—
writing or even just organizing things, such as family photo albums,
takes time and energy that younger people in the throes of child
care and career development don't necessarily have. But people's
free time could be spent doing many things. Why such a seemingly

universal urge to sum up and, in the same larger process, give back to those around them?

The entire answer to this question is not yet at hand, but research suggests that at least part of the reason can be traced to the hippocampi, those twin brain structures that are vital for memory formation and retrieval and that also help link the neocortical "thinking brain" with the limbic "feeling" brain. Recent studies by Eleanor A. Maguire and Christopher D. Frith, of the Institute of Neurology at University College, London, have discovered a pronounced difference in the activation of the hippocampi between younger and older adults when they are recalling memories from their lives. The older adults used both left and right hippocampi in the tasks, whereas the younger adults used primarily their left. These findings are consistent with those from studies of other brain regions showing greater use of both brain hemispheres in older people.

I hypothesize that using both hippocampi during recall of life events creates a richer, more vivid experience because the brain is drawing on a broader palette of resources for the task. The intuitive, holistic, and nonverbal brain skills that typically reside in the right hemisphere can bring an added benefit to the task of memory recall. Using both hippocampi may also simply make recollection itself a more vivid and pleasurable activity. I think the brain, in effect, relishes the chance to deal with autobiography in later life—and to do so using both engines, so to speak. I see autobiography and the larger processes and behaviors of summing up as a bit like chocolate to the brain in later life—a sumptuous activity.

Closely tied to the urge to organize and summarize one's life is a desire to give back in some way to family, the community, or the world at large. In my retirement study, almost everyone in their seventies is involved in some form of volunteerism, a trait that remains strong at least through the next decade. This is not simply because older people have no options for paying work. As we will see in chapter 7, older adults have more opportunities than ever to work if they want to. Nevertheless, many older people volunteer. In a national study reported by AARP in 2003, for example, 40 percent of survey participants seventy or older were involved in formal volunteer activities through some form of organization; when informal volunteer activities not in conjunction with an established organization were included, the proportion rose to 80 percent.

Formal philanthropy is common among people in the summing up phase. When participants in the retirement study are asked, "What gives you a sense of meaning and purpose in life?" one of the most common answers is making a contribution that helps others.

My wife's great-uncle, Harold Alfond, the son of poor Russian immigrants, started from next to nothing, worked hard, and eventually founded the Dexter Shoe Company in Maine. He later invented the factory outlet store and became a part-owner of the Boston Red Sox. When I interviewed him at eighty-nine, he was clearly in the summing up phase, giving back to his community and society through extensive monetary contributions and teaching about methods for effective philanthropy. He told me that "it is important to teach about how to give," and he wanted his giving "to be a model for his children, grandchildren, great-grandchildren, and other philanthropists." His four children have themselves all become philanthropists. At the time of our interview, Harold had donated over $100 million—more than half of it given since he was eighty. Expressing a sentiment common in this phase, he said, "I want this country and the world to be a better place for everyone."

The summing up process can also lead people to confront unfulfilled dreams and to bring closure to unresolved conflicts. The pressure to confront and come to terms with these matters can be very powerful in this stage of life. In the early 1960s, psychiatrist and Pulitzer Prize–winning gerontologist Robert Butler developed the concept of the "life review" and put it to therapeutic use with individuals. Butler says the life review process is "characterized by the progressive return to consciousness of past experiences and particularly the resurgence of unresolved conflicts that can be looked at again and reintegrated. If the reintegration is successful, it can give new significance and meaning to one's life."

The act of recording one's life story has gained widespread respect and acceptance among those who study or treat older adults. Psychologists have learned that reviewing one's life is part of normal aging. The review process can then lead to self-awareness and self-acceptance. Butler's pioneering work has now been taken in many directions. Life review can be done in groups or privately via writing, artwork, photography, or practically any other expressive medium. The point is that summing up, in any form, can be a stimulating, life-opening experience.

Summing Up Phase Stories

It's never too late to resolve conflicts around unfulfilled dreams and unfinished business. Sometimes resolution means realizing those dreams. Consider the following stories:

Frank Bourgin's Vindication.   Frank Bourgin's unfulfilled dream had weighed on him for more than four decades. Forty-five years earlier, when he was in his thirties, his Ph.D. dissertation manuscript was rejected by the University of Chicago. His thesis defended Franklin D. Roosevelt's then-new social programs against charges that they were unconstitutional. If he wanted to challenge the rejection, he would have to enroll again as a full-time student. But Frank was married and had a newborn child; reenrolling was not an option. He found a job in business and put his thesis in a steel box. But Frank never lost his nagging sense of injustice about the denial of his dissertation. In 1987, on the occasion of the bicentennial of the U.S. Constitution, he decided to make another effort. He mailed a summary of his thesis to the historian Arthur Schlesinger Jr., who had just published a book that made some of the same points Frank had made decades earlier.

Several weeks later, a letter arrived from Schlesinger. Frank Bourgin cried when he read it. Schlesinger praised the work as pioneering and a "strikingly original piece of work." He also said that he would urge the political science department at the University of Chicago to take a second look at it. The department did, and the following year, Frank rolled across a stage in a battery-powered scooter to receive his Ph.D. degree at age seventy-seven.

The Second Half of Life: Phases III and IV

Verdi: Bridging a Fifty-Year Gap.   Many marvel that Giuseppe Verdi was in his eightieth year when he composed his celebrated opera *Falstaff*. But why did Verdi choose to compose *Falstaff* rather than a different opera? The dynamics of the summing up phase offer one explanation. It turns out that Verdi had unfinished business that had gnawed at him for more than half a century.

When he was twenty-five, Verdi composed a comic opera— *Un Giorno di Regno* *(King for a Day)*. It opened in 1840 in the famous theater La Scala but was received so poorly that it was canceled after one performance. The failure was especially crushing because Verdi had recently lost his wife and, a year earlier, his infant son. He vowed never to write another opera, but, under the steady encouragement of La Scala's director, Verdi eventually wrote *Nebucco*, the success of which launched Verdi's decades-long career.

Fifty-five years after *King for a Day* flopped, Verdi, in his summing up phase and at the top of his field, looked back at unfinished business—his failure to compose a successful comic opera. He decided it was time to try again. The result was *Falstaff*, one of the finest operas ever written. And, in a fitting coda, the opera premiered at La Scala.

Summing Up En Masse.   Jim Grenquist, a participant in my retirement study, described to me how for the past decade a group of older adults in his hometown of Malden, Massachusetts, has been coming together to share stories about their lives. (The stories are particularly meaningful for me because my aunt, Rose Litchman, lived in Malden during the same period as the group's participants.)

The group started with six adults who had grown up in two sections of Malden—Linden and Maplewood. They knew the richness of historical and cultural stories from the 1930s and 1940s and wanted to preserve them. They found that sharing their stories with others who had grown up in the same era triggered yet more memories.

As the years passed, excitement about the project grew, and eventually more than 500 people contributed memories. One of the group's members, William T. J. "Bill" Dempsey, born in 1924, volunteered to transform their mounting oral histories into books. Several volumes were eventually published using desktop technology. The preface of the first book includes the following text, which epitomizes both the impetus and the value of the summing up phase: "This book of our memories about growing up in Linden can serve as a remembrance not only for ourselves but for our children and grandchildren; and for the many wonderful kids in the community today it can become a proud part of their heritage."

PHASE IV: ENCORE

The encore phase generally starts during one's late seventies, becoming more pronounced during the eighties and extending to the end of one's years. I see this time as a manifestation of our creatively restless brain creating an Inner Push for reflection and a desire for continuation and celebration. I use *encore* in the French sense of "again" and "still continuing." This phase is not a swan song but more like a set of variations on the myriad themes created during one's life.

Despite illness or physical limitations, during this time of life people are still driven by powerful forces, such as the desires for love, companionship, self-determination, control, and giving back. In a sense, the encore phase contains aspects of the three previous phases within it—we still reexamine our lives, still may feel a sense of liberation from previous constraints, and still may feel keenly the desire to sum up our lives and express that reflection in some form. In this late stage of life, these Inner Pushes can be expressed in surprising ways. This is a time when entirely new perspectives on life can come forth—and as much as adults in this phase are living in well-worn grooves of behavior and outlook, they are also capable of "jumping the tracks" in spontaneous and wonderful ways.

At the ages of 105 and 103, respectively, Bessie and Sarah Delaney, African American sisters who had lived their lives together, wrote *The Delaney Sisters: The First Hundred Years*, a joint autobiography that illustrates both the summing up and the encore phases. When Bessie died two years later, Sarah wrote an encore book: *On My Own*.

The encore phase is often laced with a sense of humor about the realities of mortality and the sometimes frustrating aspects of growing old. Consider the great comedian George Burns.

I interviewed Burns when he was ninety-seven for a series of public service messages on aging. When I asked how he was adapting to his advanced age, he quipped, "I now ask for my applause in advance, just in case." But applause is what he continued to receive, right to the end at 100. And he continued with the unexpected right to the end as well. When I asked him, "What does your doctor say about your smoking and drinking?" Burns didn't miss a beat: "My doctor is dead," he said.

The encore phase is all about vitality—a vitality of spirit if not always of the body. I'm reminded of the lines from Dylan Thomas's famous poem:

> Do not go gentle into that good night,
> Old age should burn and rage at close of day.

It's important to remember that even in late life, the brain retains key abilities. To be sure, some aspects of intellectual functioning decline with age, but learning is always possible, and the depth of experience encoded in old brains is irreplaceable. New dendrites, new synapses, and even new neurons continue to be created—especially when the older adult is actively engaged in activities that are physically or mentally stimulating.

Not only does the brain itself remain resilient (as long as chronic or degenerative disease doesn't interfere), but mood and emotional outlook can also remain vigorous, even in the face of serious physical problems. This is a salient feature of the encore phase—artful continuation with life despite obstacles. The very old can still experience deep pleasure and satisfaction, especially in relationships with family and significant others. Researchers at Georgia State University in Atlanta have confirmed this. In one study, scientists looked for differences in coping skills and life satisfaction among middle-aged, young-old, and oldest-old adults. The participants were healthy, socially active adults who lived in the community. No significant differences were found in eleven of twelve coping resources or in overall coping effectiveness. Measures of life satisfaction were also equivalent across the three age groups. The findings indicate that for healthy adults, the oldest old cope at least as effectively as their younger counterparts, despite the likelihood that they will face physical ills or loss of one sort or another. The relatively high levels of life satisfaction also demonstrate the psychological resilience and fortitude of people in late life.

Several other studies have confirmed these findings—even among unhealthy adults. The bottom line: people become better at adapting to their conditions as they get older. Regardless of their health status, older people typically are better prepared—in terms of both satisfaction with life and coping capacities—to face the vicissitudes of aging. In trying to understand why well-being rises or remains stable in later life—as opposed to demonstrating a progressive fall—researchers often point to findings that aging is accompanied by more positive and less negative emotions. As we saw in chapter 1, these findings may be related to the "mellowing" of the amygdalae, the brain's emotional processing centers. To these physiological changes I would add that the ever-evolving Inner Push for growth and development has a positive impact in this phase.

These trends are evident even among centenarians—people living to the age of 100 or beyond. In his study of centenarians, reported in *Living to 100*, Thomas Perls reports that "centenarians do not suffer long, gradual declines in health. About 95 percent of our centenarians are physically healthy and cognitively independent into their nineties, with low rates of mental illness and depression. . . . Centenarians are far more likely to have a near lifetime of health, followed by a quick decline before death." This is good news, especially since the fastest growing age group is 100-year-olds, and many people fear that living so long will mean living with a long period of disability before death.

One participant in Perls's study was Dirk Struik, who exemplifies the vitality that can accompany the century mark. Struik was a professor at MIT. At age 100 he traveled to his native Netherlands to lecture and visit relatives. At 101, he published an article on a new concept called ethnomathematics, and he was writing his autobiography. Professor Struik was still studying, thinking, pondering old concepts and learning new ideas when he passed away at age 106 in 2000. He and other centenarians practice instinctively what research on cognitive aging shows: learning new things and remaining mentally active are keys to mental vigor.

Celebrating the Encore Phase

Reunions and celebrations planned by families and communities can provide a vital social glue and cohesiveness in this phase. The "oldest old" often generate feelings of hope and possibility in their younger counterparts as well as a sense of fascination and wonder about life itself. This effect was part of the national appeal of television weatherman Willard Scott's regular acknowledgments of centenarians on the *Today Show*. Although now discontinued, those public acknowledgments created an ongoing celebration of old age shared by millions.

I often feel awe, respect, and curiosity about the oldest participants in my studies. I remember a moment that occurred more than ten years ago. I was asked to conduct a public interview of a centenarian at the Huffington Center on Aging in Houston. Mildred Horton had just turned 100, and she had a reputation as something of a firecracker. On the appointed evening, Mildred walked onstage before several hundred people. She wore a stylish belted blue-gray print dress with an elegant white unbuttoned sweater, using a walker that looked like an elegant, compact carriage. She sat in a chair facing me and I began the interview, first thanking her for taking time to talk with me and the audience. Well, she replied, she normally plays cards on this night, and it wasn't easy shifting around her busy social calendar, but she felt it was an important thing to do. That pretty much set the jocular tone for the evening.

"I have to ask you a question that I suspect you've been asked many times, but many people would still love to hear your response—what's it like being 100?" I asked. Without missing a beat, she quipped, "I don't notice any change at all from when I was ninety-nine."

In the course of the interview, she discussed her passion for helping others, which at that time was focused on the families and children in Bosnia suffering from the raging civil war. Throughout the evening, Mildred charmed the audience with a wit, wisdom, and wry humor that I've repeatedly found in people at the end stage of life.

Encore Stories

Several years ago at a conference on creativity and aging in Santa Fe, I attended a juried exhibition of works by older artists. A colleague told me not to miss the sculptures of Beatrice Pearse, so I sought her out. Beatrice was standing by her sculptures of owls, looking like a work of art herself, wearing a colorful smock over a long black dress. She greeted me with a welcoming smile and sparkling eyes.

"Those were good words you delivered in your talk," she said.

"Well, it's clear to me that you and your work help make my case about the prevalence and potential for creative expression with aging," I replied.

"Well, I haven't been an artist for long," she said apologetically. "I was a legal assistant all my life, and I didn't start sculpting until I was ninety-four. I don't know what I did with those first ninety-four years!"

Remarkably, Beatrice had such drastically impaired vision that she was legally blind. She told me that she grew up on a farm and was fascinated by the sight and sound of barn owls. Those images were etched indelibly in her mind. When she decided, on a whim, to try her hand at sculpture at the age of ninety-four, she discovered a love for the feel and texture of clay. She remembered the owls and found that she could see them clearly in her mind's eye and was able to transfer that clarity to the sculptures. She soon started exhibiting her work and received positive feedback and recognition.

About four years after meeting her, I received a letter from Beatrice's family. They told me she had died but that she had continued sculpting until a week before her death. Beatrice's story reminds me of Beethoven, who composed some of his most monumental pieces after he had become deaf. Beatrice lost her eyesight, but she continued to have a great inner vision of barn owls and other farm animals. Beatrice's story demonstrates that the inevitable losses that accompany aging need not provoke crisis or withdrawal from life.

The encore phase qualities of continuation and celebration despite loss are also illustrated in a story from my own life. My mother, Lillian Cohen, was eighty-four when my father died with Alzheimer's disease after a marriage of nearly sixty years. Three years later, she decided to move to a private apartment in an assisted living facility in the town where my brother lived. She wanted to bring with her the old piano she had grown up with, although she hadn't played it during or since the time my father was very ill. The piano weighed about 1,200 pounds, but it was successfully moved into the apartment.

Her personal encore was a return to playing the piano after more than five years. Her neighbors loved the sound of music coming through the walls of her apartment, and one of the assistants at her building asked if she would play at her wedding. My mother, recharged by the invitation, happily accepted.

My mother later had a stroke that made it difficult to carry out many of her previous activities and robbed her of some—but not all—of the memories she cherished. Nonetheless, she remained as active as she could. She was also still perfectly capable of enjoying the pleasures of life. In the same month of her stroke, her first great-grandchild was born, a baby girl named Ruby, which filled my mother with delight. Not long after that, our clan gathered to celebrate my mother's ninetieth birthday. It was exactly the kind of family gathering that is typical of the encore phase, a time to bond, recall shared memories, enjoy the pleasure of each other's company, and pay respect to the family matriarch. During the party, my mother held Ruby in her arms for an hour, and Ruby seemed perfectly content to be there.

Anna's Story

Anna Franklin had just turned 100. She lived in a public housing complex, one of several places I visited in the course of my research on aging. She had heard that I was a psychiatrist and asked if I could pay her a visit. Curious about her request, I went to her apartment one day and knocked on the door.

She came to the door without a walker or cane and smiled as she greeted me. After some small talk over tea and some cake she had just baked, I asked her why she wanted to see me. A sweet, mischievous smile spread across her face.

"Well, I'm a hundred years old and I've never seen a psychiatrist," she said. "Frankly I just wondered what you were like. Tell me what psychiatrists do."

"First they ask you if anything is bothering you, and then they ask you to tell them about your life in order to get to better know you and help you," I replied. "Is anything bothering you?"

Anna smiled again and said "Not really," she said. "At 100 you have a number of aches and pains and medications, but I live by myself and get about pretty well. If I go out, my granddaughter or a friend might accompany me. I'm blessed and have a loving family."

We talked for more than an hour about her life, growing up in the segregated South, helping her husband run a small restaurant, raising her four children, and watching as her grandchildren in turn grew up, went to college, and made their way in life. She still enjoyed cooking and coming up with new recipes, and she liked to sew and knit. At one point she took a large photo album from one of her shelves and handed it to me. She had eleven grandchildren, twenty-six great-grandchildren, and three great-great-grandchildren.

I asked if she or any of the family was helping her to write down or record the extraordinary span of her family history. She took another book from the same shelf. Prodded and helped by one of her granddaughters, she had written her memoirs a decade ago. She had also been interviewed by one of her great-grandsons for an oral history project.

"What's next?" I asked her.

She returned to the same shelf and pulled down yet another book. It was a loose-leaf notebook titled "Anna Franklin's Greatest Recipes: The First Hundred Years."

"Well, I have a lot to do," she said. "I'm always adding to this."

Before I left, I asked what she thought about her first meeting with a psychiatrist. She laughed and said, "You're not all crazy after all." She then gave me a big hug.

I left Anna Franklin's apartment exhilarated by her vitality and warmth. She personified the encore phase of life!

THE FOUR PHASES: SUMMARY

We've now explored the four phases of older life: midlife reevaluation, liberation, summing up, and encore. These phases represent the combination of neurological, cognitive, and emotional development over time, which is a manifestation of the continuing Inner Push. As in earlier phases of life, our later years are imbued with drives for new perspectives, change, and new forms of creative expression. The phases emerge in a loose progression, commonly but not always separated by time; sometimes they overlap, intersect, or synergize. Each individual experiences his or her own sequence, pattern, and outcomes of the four phases. And the positive influence of each phase reflects the ongoing dynamism and resilience of the human brain.

We see the trajectory and force of human development in a baby who strives to grasp, crawl, and communicate; a toddler who wants to climb the highest stairway; and the young child who creatively turns a bed into a sailing ship and pillows into fortresses. We see it in the adolescent's persistent experimentation and drive for autonomy. We see it as well in the creative pursuits of adults of all ages who want to figuratively or literally climb mountains and reach for the stars. We see it in centenarians confronting the upper limits of our life span with their own kind of creativity and courage. At work in all these stages is that Inner Push built into all of us— urging us toward new paths, challenges, change, and creativity. Ultimately, these developmental aspects of aging contribute not only to our individual growth and well-being but to the health and survival of our species.

5

Cognition, Memory, and Wisdom

The doors of wisdom are never shut.
                    —Benjamin Franklin

ON A TABLE IN A HIGH-CEILINGED ROOM in the State Museum of Georgia in the former Soviet Republic, sits a hollow-eyed skull. The skull is not as mute as it may appear. To the anthropologists who unearthed it, the skull whispers messages about the importance of the elderly in ancient protohuman societies. This skull is the earliest known evidence of the emergence roughly 1.8 million years ago of a powerful new force in the evolution of human beings: wisdom.

The anthropologists at the museum refer to the skull as "the old man," even though they estimate that the hominid—a member of the *Homo erectus* species—was about forty when he died. Indeed, all the other skulls found in the same stratum of soil appear much smoother, with small, graceful features and intact teeth, indicating a younger age at death. Reaching forty in that time was probably equivalent to reaching the century mark today. But the stunning thing about "the old man" is this: Not only does he have no teeth, but the tooth sockets are smooth, filled in with bone that grew over the spaces. This bone regrowth shows that "the old man" lived several years after his teeth fell out. At a time when hominids were very likely as often prey as predators, the most plausible explanation for this curious feature is that he was helped by his fellows—he was fed and cared for. And this, in turn, means that he was valued.

We cannot know what this old man contributed to his community, but most plausibly he was valuable because of what he *knew*. Although pinpointing the origins of language will never be an exact science (sounds leave no fossils), available evidence about brain size and the structure of rib cages leads some anthropologists to suggest that *Homo erectus*, and maybe even the earlier *Homo habilis*, were capable of producing some form of language. It may have been the invention and use of language, in fact, that was the decisive factor in the success of *Homo erectus* over the other hominid species alive at the same time, such as the Neanderthals. In any case, the ability to transmit knowledge from one generation to another must have been an enormous asset and, as a consequence, would have raised the value of those members of a community who had accumulated the most knowledge—the elders—and were able to transmit it to others.

As the body of human knowledge grew over the ages, and as the social and cultural life of our species became more complex, the value of older adults increased as well. This is why natural selection has favored a relatively long human life span despite the fact that the female reproductive potential typically ends in the late forties. Clearly, when it comes to human beings, the capacities to teach, to transmit wisdom and skills, and to serve as a repository for the culture are just as important as the capacity to reproduce. The complexity of today's global society and the variety of skills required to master those complexities only amplify the importance of the older adults among us.

WISDOM AND POSTFORMAL THINKING

What exactly is wisdom, and how does it develop? One standard definition is that wisdom consists of "making the best use of available knowledge." This rather utilitarian approach implies that wisdom requires specific knowledge as well as a broad understanding of the context in which that knowledge can be put to use. But this definition isn't completely satisfying. For most people, wisdom also connotes a perspective that supports the long-term common good over the short-term good for an individual. Insights and acts that many people agree are wise tend to be grounded in past experience or history and yet can anticipate likely future consequences. Wise acts, in other words, look both backward and forward. Wisdom is also generally understood to be informed by multiple forms of intelligence—reason, intuition, heart, and spirit. It is fundamentally the manifestation of developmental intelligence—a mature integration of thinking skills, emotional intelligence, judgment, social skills, and life experience.

Contemporary psychology has elucidated a key component of a more advanced thinking style (cognition) that is an important ingredient of wisdom: postformal thought. I mentioned this concept in chapter 2 because postformal thought contributes significantly to our developmental intelligence. Formal thinking emphasizes pure logic in problem solving and is best suited for well-defined problems with clear rules of operation, such as in mathematics and the "hard" sciences. Jean Piaget, one of the founders of the field of cognitive development, believed that this quality of "pure thinking" peaked in adolescence and young adulthood. He thought that being "mature" meant thinking like a scientist.

But many people now think this view is too limited. The real world is not as neatly defined as mathematics. Rules are not always clear, and knowledge is not always absolute. The concept of postformal thought was created to describe the more subtle, flexible, and insightful thinking styles that can only develop over time. *Postformal* thinking is valuable for ill-defined, ambiguous problems for which more than one solution is possible. It focuses more on what is relative rather than on what is absolute in nature and as much on problem identification as on problem solving. Most of the knotty ethical dilemmas facing us today fall into this category. Using embryonic stem cells to cure disease, for example, involves weighing competing value systems, clarifying the issues at stake, and considering a myriad of solutions in the search for common ground.

Recall that postformal thinking involves three types of reasoning—relativistic thinking, dualistic thinking, and systematic thinking, as elaborated in chapter 2. Each of these modes of thought results from the ongoing cognitive development as we grow, learn, and gain experience. Postformal thinking fits into the larger concept of developmental intelligence because it allows us to better integrate our emotions and reason when we're trying to solve a problem. Our increased ability over time to draw on a broad set of mental skills is probably related to the increasing bilateral involvement of the right and left brain hemispheres that is associated with the maturing brain.

Many of us in our second half of life remember times in our youth when we felt we knew a great deal but somehow couldn't put our understanding together to resolve a difficult issue. By midlife, however, we often have gained enough insight to deal effectively with problems that previously bedeviled us. For example, most high school students know the basics about alcohol: that it acts primarily as a central nervous system depressant, that is can lower inhibitions, and that it can make you sick if you drink too much. That information alone, however, is seldom enough to allow an individual to use alcohol wisely (or choose not to use it at all). Alcohol use is complicated by an array of variables, such as one's genetic predisposition to addiction, one's idiosyncratic metabolism, the fact that alcohol's effects are strongly related to the setting in which it is used, and that it exerts qualitatively different effects over the course of inebriation. With experience, one learns about these subtleties and about the specifics of one's own body and life circumstances, which (in the absence of outright addiction) can lead to wise decisions about alcohol use. Logic alone is an insufficient guide—you have to draw on postformal thinking.

A historical example of postformal thinking in action comes from the experience of a young nineteenth-century naturalist.

Between the ages of twenty-two and twenty-seven, this man traveled the world collecting thousands of plant and animal specimens and recording his observations in dozens of notebooks. Although he made interesting discoveries of new species, he struggled to see a bigger picture. It took him twenty-three years of in-depth thought, research, and correspondence with others before the pieces of the puzzle fell into place in a way he felt was suitable for publication. At the age of fifty, his book was published to both acclaim and condemnation. The man was Charles Darwin, and the book that took so long to create was *On the Origin of Species*.

POSTFORMAL THINKING AND THE PHASES OF OLDER AGE

In phase I, midlife reevaluation, our emerging capacities for post-formal thinking can both accentuate and help solve the feelings of ambivalence that are common at this time of life. The postformal abilities to contemplate more than one answer to a problem, to consider contradictory solutions to life's challenges, and to recognize how much in life is relative are exactly the tools we need in this phase. We can begin to raise new questions, consider alternatives before jumping to new solutions, and make decisions based on a tighter integration of how we think and feel. This is the underlying psychological engine driving the usually constructive (albeit emotionally challenging) midlife reevaluation phase. At the same time, our changing inner psychological climate, influenced by postformal thought, gives us greater respect for intuitive feelings.

The postformal merging of "heart" and "mind" is also helpful in the liberation phase, as illustrated by the following story.

Marilyn Andrews always planned to retire with her husband when she was sixty-five. But when she was sixty-two, her husband died unexpectedly. Marilyn had worked as the administrative assistant to a lawyer for more than twenty years. When she was sixty-four, the lawyer retired and closed his office. Now a widow, without children, Marilyn needed another job. Despite a glowing letter of recommendation and clear expertise, she was turned down by several potential employers when she told them she was sixty-four. They said they wanted someone who would stay on the job for ten to fifteen years.

All her life, Marilyn had been a straight arrow, always following the rules. But now she realized her age was working against her even though she looked much younger than her years. "What if I just lie about my age?" she wondered. The thought of getting away with it made her chuckle. "Well, why not?" she said to herself. "What's the worst that could happen? They could fire me."

Marilyn felt energized by her plan. She bought a new outfit that made her look younger yet, and interviewed for another job. Like the others, the interviewers were impressed with her résumé (which omitted her age) and her experience. When they asked her how long she hoped to be with them, she said fifteen years, until she could retire at the age of sixty-five. "I listened to my head, but I guess I went with my heart," she said. She got the job and kept her word, retiring fifteen years later at the age of eighty. "What an adventure it was!" she said.

I see this as the liberation-phase Inner Push at work, combined with an evolving postformal thinking ability, nudging her to take a risk and a path of action previously foreign to her.

The story of Sam Sheldon, seventy-eight, illustrates cognitive growth in the summing up phase. Sam, who participated in one of my studies, told me that he wanted to try writing fiction. He decided to exercise his writing abilities by writing a memoir, which is a classic indicator of the summing up phase. "I know it's not fiction," he said with a chuckle, "but some of my real experiences have been stranger than fiction."

In the middle of his memoir project, he had a heart attack. He recovered relatively quickly, but the experience pushed him to consider moving from the house in which he had lived for forty-five years. He had three options: a smaller house; a condo in a building with a range of age groups; or a retirement community with various health care options if he needed them. Instead of just mulling over these potential options, he had a great idea: he would develop three written scenarios describing what life might be like for him in each setting. It was an excellent example of wisdom—of mixing objective and subjective problem-solving skills. In the end he decided to move into the condo, although he realized that the assisted living option might make sense if his health worsened. "When I started to really visualize my days in each scenario," he said, "I found there was more to write about when I put myself in the condominium with a mix of people than in either of the other two options."

Postformal thinking and wisdom are also often evident in the encore phase. I met Elinor Frank, age ninety-six, during a visit to an assisted living facility. She said that a week earlier, another resident had gotten annoyed with her. "You're acting silly," the resident said to Elinor. "Why don't you act your age?" Instead of getting angry in return, Elinor used a strategy typical of postformal thinking: answering a question with another question that reframes the issue in larger terms.

"How should I act my age?" she asked, with encore phase spirit. "Tell me how I should act when I'm eighty or ninety or 100. Isn't it good to be a little unpredictable once in a while, so people don't take you for granted?"

WIRING WISDOM

The maturation of our thinking abilities as well as our overall growth in developmental intelligence depend critically on behind-the-scenes brain changes. Our brains never lose the ability to learn by forming new synapses, dendrites, and even entirely new brain cells. These fundamental capacities, as well as the aging brain's apparent ability to recruit new regions of the other hemisphere for specific tasks, more than make up for the real—but gradual—losses in the speed of signal transmission or the loss of brain cells. But yet another aspect of brain development provides reason for optimism.

If you sliced into a section of brain, you would see a relatively thin top layer of gray-colored tissue covering the bulk of the brain tissue, which looks white. The gray matter is composed primarily of the actual bodies and dendrites of brain cells. Billions of these cells endow us with our many mental functions, such as perception, language, and self-awareness, as well as bodily control. The white matter is composed mostly of the long tail-like extensions of the brain cells—axons—which carry messages to other cells in the brain or body. Axons are like high-bandwidth cables capable of carrying a signal over long distances. The white matter, in other words, is like the Internet wiring connecting all of the "users," which are the cells and networks of cells in the gray matter.

Each axon is surrounded by a fatty substance called myelin, which acts like an electrical insulator, greatly speeding up signal transmission. It's the myelin that makes the white matter white. Myelin is added to axons continuously through the first forty years of life, peaking in volume around age fifty, but continuing to form at a slower pace to the end of life. The practical upshot of this ongoing construction is better coordination between the brain's many modules, more effective integration of the brain's hemispheres, and more efficient signal transmission throughout the brain, all of which support the more flexible and nuanced thinking characterized by postformal thought and wisdom in the second half of life.

A number of studies support this point, showing that many of our intellectual and cognitive abilities peak not in young adulthood but in midlife or beyond. Psychologists Sherry Willis and K. Warner Schaie, of the Seattle Longitudinal Study, have followed a group of men and women since 1956. They find that subjects at midlife score higher on almost every measure of cognitive functioning than they did when they were twenty-five—verbal and numerical ability, reasoning, and verbal memory all improve.

HELEN'S STORY

In my book about creativity in later life, *The Creative Age*, I drew a distinction between "big C" creativity—those great and enduring works of art or invention—and "little C" creativity, which are small, often spontaneous acts of novelty in everyday life—ordering a pizza for delivery in order to hitch a ride to a dinner date, for example, as my in-laws did in chapter 1. I think the same concept applies to wisdom. There are "big W" types of wisdom, such as the democratic and humanist ideas imbedded in the U.S. Constitution, and "little W" types of wisdom—the deft solution to a playground tussle or sage advice about relationships offered to an upset teenage girl. The story of Helen Herndon, a participant in one of my studies, provides an illustration of this "little W" type of wisdom that arose from her maturing developmental intelligence.

At sixty-seven, Helen found herself alone in a large house, her two children married and living hundreds of miles away, her husband dead after a battle with leukemia. For more than a year she existed in the twilight world of bereavement, unfocused and lethargic. The responsibilities of maintaining the house without her husband's help overwhelmed her, but she lacked the self-confidence to find a solution.

The youngest of five children, she had grown up being taken care of. "I seemed to just assume that if something was important, I couldn't do it by myself," she told me.

But as the months passed, Helen—almost in spite of herself—grew increasingly impatient with her own passivity and inertia. When one of her children suggested that she join a widows group, she followed up on it. Her decision to act sprang from deep internal stirrings for connection, meaning, purpose, and sharing. It proved a wise decision indeed.

Before she was married, Helen had earned a master's degree in special education. She never taught, however, because after graduation she met her husband and they soon married and had children. Although her lack of self-confidence prevented her from seeking even part-time work in the years to come, she kept up with the field, reading the professional literature and periodically attending seminars and conferences about childhood reading disorders.

Now, fueled by the Inner Push, Helen sought ways to tap that expertise. She began with a decisive step to simplify her life: she sold the family home and moved into a condominium complex in which several other members of her widows group lived. She started thinking seriously about becoming a reading tutor and talked about it in her group.

"These were new feelings for me," she said.

Aware of Helen's skills, a member of her group one day asked if Helen would evaluate a granddaughter with reading problems. Helen met with the little girl, who took an instant liking to her. The tutoring went well, and word of mouth brought Helen other clients.

One day Helen had the creative thought that it would be constructive to hold some of the tutoring sessions at the local library, where the children would be surrounded by books. During one of her afternoon sessions, a reporter from the local newspaper saw her and interviewed her for an article about the library as a community resource. The article and accompanying picture gave Helen some valuable publicity. Her self-confidence blossomed along with a new-found social savvy. She created a brochure about her work and distributed it to nearby elementary schools. Again, she was both surprised and pleased with her energy and confidence. "I almost didn't recognize myself," she said, smiling.

Helen now has a busy schedule as a tutor, is fully engaged with her life and her community, and is serving as a valuable source of knowledge and wisdom for both the children and adults around her. She is a marvelous example of someone in the liberation phase of late-life development, but she also embodies someone who was able to harness her emerging wisdom to create positive new opportunities in her life.

MEMORY: THE FOUNDATION OF WISDOM

Earlier I described wisdom as deep knowledge used for the highest good. Making wise choices usually involves drawing on both the logical and the intuitive, the right and left hemispheres, the head and the heart. I believe that the changes and continuing growth of the brain in later life amply support our capacity for wisdom and that, as the quote from Ben Franklin at the start of this chapter suggests, this capacity never runs dry.

I want to look a bit deeper into this phenomenon. In particular, I want to explore the brain process that is fundamental to wisdom and, indeed, to all of our cognitive functioning: memory. To think, to decide, and to act wisely require full access to our vast memory library—old memories and new, verbal and nonverbal memories, emotional and intellectual memories. As anyone who has seen someone in the grip of Alzheimer's disease knows, without memory, our very identity disappears. I have found that many older adults fear losing their memory even more than they fear death. Most of us understand that a life without memory is a ghost life at best.

Fortunately, neuroscientific research offers some comforting findings. While the overall processing speed of brain cells and certain specific types of memory do decline gradually with age, other forms of memory robustly resist degradation. Global memory loss is *not* an inevitable consequence of aging, and many steps can be taken to stabilize or improve memory as we grow old. And here's another little-known fact: Unlike a computer's hard drive, our brains have no known limits for memory storage. In other words, just because you're old, that doesn't mean you've "used up" your brain's memory capacity.

To see more clearly why this is so, you need to understand a bit about the brain's mind-boggling circuitry. As we saw in chapter 1, the brain is composed of billions of individual neurons, and each of these neurons has thousands of dendrites and dendritic spines that connect them to other neurons. The number of potential ways these dendrites can connect with each other is so vast as to be nearly inconceivable. And, since memories reside in these patterns of connectivity, the limits on memory are logistical, not fundamental. We are limited only by the time we have in life for learning—our brains (as long as they're healthy) could contain many lifetimes of information.

To better understand this, picture a crowded football stadium. In one section, 200 people are holding cards painted red on one side, black on the other. On the command of a leader, everybody holds their card above their head, showing either red or black. With practice and coordination, these 200 people can spell out an unbelievable number of words or visual patterns—"Go Team" being just one of millions. This is a bit how memory works. Experience rearranges the connections between brain cells, allowing a given neural net to "remember" a vast number of different firing patterns. Those patterns might provide the code for a letter of text, a face, a musical phrase, or the scent of a fresh-cut lime. The bottom line: Our brains contain not 200 neurons, but *billions* of neurons, which is why the brain's memory capacity is essentially limitless. This fundamentally hopeful message about aging is only just beginning to reach the public consciousness.

Other aspects of memory deserve attention as well. Most people know from experience about two broad types of memory: short-term and long-term. But we now know that human memory is a good deal more complicated than this. Today, scientists use the term "working memory" for what most people call "short-term memory." Working memory is like a mental desktop where you temporarily keep information required for a task at hand. This is where you store a phone number while you dial, or street directions someone gives you. It is also how you can grasp the meaning of a sentence—your brain automatically holds the first part of a long sentence in working memory so it is available for processing by the time you reach the end of the sentence. Working memory can store images as well as words and numbers. Artists constantly use their visual working memory to retain images long enough to transfer them into whatever medium they are working with.

Working memory is not particularly powerful: it cannot store large amounts of information, and information typically resides on the "desktop" only briefly. In order for a memory to be captured in a more permanent way, it must be transferred to a different part of the brain and linked to networks of memory associations laid down by previous experience and learning. This is "long-term" memory. It is composed of millions upon millions of constellations of brain cells, the connections between which encode everything from the scent of the glue you used in kindergarten to the image of a lover's face to the muscle memories for riding a bicycle. Long-term memory is really a bit of a misnomer because it implies a single storage location in the brain. In fact, long-term memories are distributed all across the brain. This is not only a very efficient system for the brain (visual memories being stored in the visual perception areas, for example), it also helps protect the brain from catastrophic memory loss. If all our memory eggs were stored in a single basket, a stroke, a blow to the head, or other damage could be devastating. By distributing our memories, we can lose some types of memory while retaining others.

Long-term memory comes in two broad types: declarative and procedural. As its name suggests, procedural memory is where we store "how-to" information. It is largely a nonverbal type of memory and is experienced as our "body memory" for riding a bike, shooting a basketball, playing a musical instrument, or driving a car.

Declarative memories can be consciously called upon and "talked about" in words or drawings. Some declarative memories, called semantic memories, are facts about the world: the names of things, sports statistics, mathematical formulas, letters of the alphabet . . . the list is nearly endless. Other declarative memories are more image based and represent scenes from our experience, usually from a specific place at a specific time: where we were when we heard about the September 11, 2001, terrorist attacks, what our grade school looked like, and the color of a prom dress. This "episodic" declarative memory is also mind-bogglingly rich. Although memories can never be as detailed as present reality, our storehouse of life memories is impressively large. Indeed, it is larger than we realize because it is constructed in vast interconnected webs of associations. Most people have had the experience of suddenly remembering a hitherto forgotten scene from the past, triggered by a smell, sound, or sight.

It's important to understand, though, that the brain doesn't store a photographic record of our life that we can access at will. That would tax even the phenomenal capacities our brains do have and would fill our heads with a vast amount of useless information. We only remember what stands out, what is unusual, personally significant, unpredicted, or what we intentionally memorize. Most of our stream of consciousness—the aspects of the external or internal world that we are paying attention to—passes through our brains unpreserved.

Understanding more about memory can help us make better use of it. For example, because memory is a phenomenon of *association*, we are better able to retrieve specifics that are connected somehow to other things we have learned. A person's name, for example, is usually connected to other details about the person, such as their face, occupation, and where we usually see them. If you forget someone's name, you can sometimes bring it forth by "tickling" the neural nets connected to the one representing the name by focusing on related bits of information. Say, for example, you're trying to remember the name of a female friend; you can picture her children, where she works, or when you last saw her, all of which activate the neural nets that represent your friend, including her name, which, more often than not, will pop up after a moment or two. Memories that are well integrated into the context of other memories are more resistant to degradation than memories of specific details or facts not linked to other memories.

Memory is an endlessly fascinating phenomenon and is very much at the center of current research. Some recent studies, for example, have found that although working memory and the episodic form of declarative memory can degrade over time, semantic and procedural memories are quite stable. Often these types of memories even resist the ravages of Alzheimer's disease. More fundamentally, however, the capacity to learn never dies, and many studies show that our ability to retrieve and manipulate existing memories as well as our ability to lay down new memories can be improved with mental exercise. You can enlist a wide range of simple strategies, such as jotting things down in writing, carrying a small pencil and notebook, or using mnemonic tricks to augment your more vulnerable working memory.

Many older adults, especially those in the summing up phase, also bolster their memory systems by organizing, articulating, and collecting the visual and written records of their lives. Memories can be harvested in many ways—written memoirs, family scrapbooks, genealogical files, photo albums, or, as is becoming increasingly common, digital collections of photos, text, and even audio or video recordings. My father, Ben Cohen, used a scrapbook.

I was chatting with my father during a visit when he was seventy-seven. He had read about some of my work that morning in the *Boston Globe*. With a twinkle in his eye, he said, "Did you know that I've been in the *Globe* too?"

I was surprised and said I wished I could see the article.

"Well, you can," he replied, pulling out a thick scrapbook I didn't know he had. It turned out he had just finished pulling together notes, clippings, and photos from his life. We flipped through the book to an advertisement in the *Globe* dated December 31, 1933. It was of a handsome young Navy sailor under the headline "Navy Again Asking for Recruits Seeking a Career." Below the photo was the caption "Johnny Haultight," but in fact it was my father, then in the Navy, who served as the Navy's version of "GI Joe" in this recruitment campaign.

My dad's scrapbook became even more important a few years later, when he began to show symptoms of Alzheimer's disease. The scrapbook provided a wealth of material to help jog his failing memories.

FLUID INTELLIGENCE VS. CRYSTALLIZED INTELLIGENCE

Related to the different modes of memory are the important concepts of fluid intelligence and crystallized intelligence. Fluid intelligence is on-the-spot reasoning ability—a kind of raw mental agility that doesn't depend completely on prior learning. It includes the speed with which information can be analyzed as well as attention and memory capacity. This is the type of native intelligence that IQ tests strive (not always successfully) to measure. Crystallized intelligence, on the other hand, is accumulated information and vocabulary acquired from school and everyday life. It also encompasses the application of skills and knowledge to solving problems.

Many studies have shown that fluid intelligence slowly declines with age, whereas crystallized intelligence often improves or expands. Many people continue to gain expertise and skills in particular areas throughout life. The great historian Arnold Toynbee was referring to crystallized intelligence when he remarked, at age seventy-seven, that "my reward for having reached my present age is that this has given me time to carry out more than the whole of my original agenda; and an historian's work is of the kind in which time is a necessary condition for achievement."

Lessons from the Oldest Elder

I'd like to end this chapter by reflecting on one of the most remarkable older adults of the twentieth century: Jeanne Louise Calment. Madame Calment was born in France in 1875 and lived to be 122, making her the oldest human being for which we have unequivocal documentation. She also embodied the truth that cognitive decline is not inevitable and that wisdom is often the golden fruit of age. Her mind remained sharp to the end, although around the age of 115, her eyesight, hearing, and mobility declined rapidly. The French media regularly referred to her as the "doyenne of humanity," and throughout the 1980s and 1990s she was a regular fixture in the press (she died on August 4, 1997). Her quick wit and pungent sense of humor made her ever popular. Once, asked about the effects of aging, she quipped, "I've only one wrinkle and I am sitting on it."

She ate and drank whatever she wanted—continuing her tradition of daily glasses of port until the end, although by then she had reduced her intake of chocolate from her earlier ration—she claimed—of two pounds a week. Some of her charm can be gleaned from the following quotations, all taken from interviews conducted after her 110th birthday:

"I'm a normal woman."

"I am brave, and I'm afraid of nothing."

"I took pleasure when I could. I acted clearly and morally and without regret. I'm very lucky."

"Wine, I'm in love with that."

"I have a stomach like an ostrich."

"I dream, I think, I go over my life . . . I never get bored."

At the age of ninety, she signed a contract—a common one in France—to sell her condominium apartment *en viager* ("in life annuity") to lawyer François Raffray, then forty-seven. In the United States, this is called a "reverse mortgage." Raffray agreed to pay a monthly sum until she passed away. At that time, the value of the apartment was worth ten years of payment. Unfortunately for Raffray, not only did Calment survive for more than thirty years, but he died first, at the age of seventy-seven. His wife had to continue the payments. In her later years, Calment lived mostly off this income.

Meanwhile, she took up fencing at eighty-five and still rode a bicycle at 100. At 115, someone at her birthday party said to her, "Until next year, perhaps," to which she retorted, "I don't see why not! You don't look so bad to me."

At 120 she gave up smoking, and her doctor said her abstinence was due to pride rather than health—her vision was too poor to safely light a cigarette herself, and she hated asking others to do it for her. And at 121, in an encore to her full life, she recorded a CD, *Time Mistress*, in which, accompanied by a musical score, she reminisced about her life.

Few of us will reach 120, but I think we can all aspire to the *joie de vivre* that Madame Calment epitomized. Her attitude stands in sharp contrast to the cartoons, myths, and misconceptions that saturate our culture today.

6

Cultivating Social Intelligence

Agreeable society is the first essential in constituting the happiness and, of course, the value of our existence.
—Thomas Jefferson

JOHN AND NINA HENDERSON WERE STILL ADJUSTING to their empty
nest. Their son was married, with three children, and lived on the
West Coast. Their daughter and her two children lived nearby, but
they didn't see her or the grandchildren as much as they would
have liked.

Then their daughter got divorced, and the Hendersons
watched as she struggled to keep up her career while raising her
eight-year-old boy and six-year-old girl. One day, after listening
to their daughter complain about how much trouble her son was
having with math, they suggested—tentatively—that they move
in to help. They had been thinking about downsizing from their

three-level house anyway, and this option looked like a "win-win" solution. To their relief, their daughter enthusiastically agreed.

Mrs. Henderson, who was a retired math teacher, began tutoring her grandson, and his performance in school improved. Mr. Henderson, an avid gardener, spent many hours with the grandchildren planting and cultivating vegetables. The Hendersons now enjoy their life as a three-generation family, and their daughter finds their presence invaluable and enriching.

Such three-generation households are uncommon in the United States, accounting for only about 3 percent of all households in 2003. I'm not suggesting that this type of arrangement is right for everyone; there are undoubtedly plenty of empty nesters who fill the social-emotional vacuum in other ways and actually *like* the extra space in the nest. My point is more general: Social intergenerational contact can be stimulating and deeply rewarding, whether between family members or strangers. And such interactions stimulate our "social intelligence," our capacity to sustain existing relationships and build new ones. Social intelligence, like developmental intelligence, of which it is a part, usually improves with age.

I have heard many stories like the Hendersons' over the years. Often it's not that older adults actually move in with their children and their families, but that they move nearby so that they can more fully participate in family life. In crisis situations, it is increasingly common for grandparents to help raise their grandchildren. This trend supports the idea that older adults provide key survival advantages to the species, which is one reason nature has endowed humans with life spans that stretch far beyond the years of maximum fertility. (Human beings apparently aren't the only mammals to which this pattern applies. Research on bottlenose dolphins and pilot whales, for instance, shows that postreproductive members of a group babysit, guard, and even breast-feed their grandchildren.)

A FAULTY EARLY THEORY

Views about the importance of social relationships in the second half of life were knocked off track in the early years of gerontological research. In 1961, two prominent University of Chicago researchers, Elaine Cumming and William Henry, presented their "disengagement theory." In *Growing Old: The Process of Disengagement*, they challenged the idea that people can be satisfied and happy in old age only if they remain active and involved. Cumming and Henry argued that normal aging involves a natural and inevitable withdrawal or disengagement, "resulting in decreasing interaction between an aging person and others in the social system he belongs to."

To their credit, Cumming and Henry didn't just make this stuff up—they were reputable researchers who were basing their ideas on what they observed in the field. The problem was that they weren't necessarily observing *healthy* older adults, and they focused exclusively on one potential aspect of behavior—social withdrawal—without looking at all the other behaviors, such as the clear need everyone has for social interactions. Compounding this problem were later popularizations and misinterpretations of the theory that made it sound as though it was normal for old people to sit alone in their rooms tuning out the world. Nothing could be further from the truth!

Other researchers, sensing the inadequacies of disengagement theory, conducted their own studies and drew the opposite conclusions. Robert Havighurst, one of the collaborators on the Kansas City Study of Adult Life, formalized what he called "activity theory." He argued that the psychological and social needs of the elderly were no different from those of the middle aged and that it was neither normal nor natural for older people to become isolated and withdrawn. When they do, it is often due to events beyond their control, such as poor health or the loss of close relatives. The research done in the decades since Havighurst's pioneering work supports his ideas and casts doubt on the disengagement theory. Research on activity in the elderly included physiological studies of the brain that demonstrated the beneficial neurological effects of social stimulation. But it took a long time to recover from the wrong turn taken by both science and society in understanding the true needs and potential of older persons.

A STORY OF SOCIAL ENGAGEMENT

Donal McLaughlin was seventy when he read an article in the *Washington Post* that both amused and inspired him. An elderly giraffe named Victor in the Marwell Park Zoo south of London had died *in flagrante delicto*. As a tribute to Victor, McLaughlin decided to create a society encouraging older adults to remain active in life. He called the group the Victor Invictus Society in the spirit of William Ernest Henley's nineteenth-century poem "Invictus," which includes the lines "I'm in possession of my fate" and "I will not be conquered." The society's slogan was "never give up," and its membership grew to more than a thousand members worldwide. All proceeds from membership dues were donated to the Marwell Park Zoo.

That was in 1977. Since then, McLaughlin has ably lived up to his society's motto. In his early nineties, he organized The Society of the Porcupine in his hometown of Garrett Park, Maryland. An architect and draftsman by training (he headed the design team for the United Nations emblem), he created membership cards featuring a determined porcupine pointing to a display that reads "Don't Tread On Me" and "The Quill Is Mightier than the Sword." He created the society to defend the "spirit of the small town in America."

Deeply moved by the September 11 attacks, McLaughlin, at the age of ninety-six, entered the World Trade Center memorial competition with a beautiful landscape design to commemorate the Twin Towers. He envisioned a quiet sea of grass with a pair of oaks in the original buildings' footprints and other plantings throughout. Here is his description:

> Curling pedestrian paths ramble throughout the entire acreage leading in and out of approximately 30 cul-de-sacs. Each cul-de-sac provides space for meditation or quiet reflection on engraved quotations from great men and women of history as well as from the world's major religions.

In addition, he proposed to honor each victim with memorial plaques placed throughout the labyrinths. Survivors and family members would be given acorns of the same species as the ones planted in the memorial, and they would be encouraged to plant them to create a living memorial to September 11.

McLaughlin's design was not accepted, but that hardly discouraged him. As of this writing, he is still active in causes and organizations, such as one fighting the proposed reforms to the Social Security program. As he approached his ninety-eighth birthday, he was asked if he had any advice to offer. This "keeper of the culture" responded: "Find a cause and leave a footprint in life for people to follow."

SOCIAL AWARENESS AND THE FOUR PHASES

Donal McLaughlin illustrates the truth that social connection and active engagement are as vital in one's later years as they are earlier in life. Maturing social intelligence is one of the elements of developmental intelligence, which, as we've seen, continues to expand as we move through the phases of later life. During the midlife reevaluation phase, postformal thinking gives us greater social and emotional resiliency, reduces our tendency to judge harshly or quickly, and can result in a more forgiving attitude toward others. During the liberation phase, a new sense of freedom and comfort with ourselves can allow us to be more socially bold and outgoing, which can spark new relationships.

The forces at work in the summing up phase can push us to connect with our larger community, with our past, and with our emerging role as a resource for history, expertise, and wisdom. Social connections play a key role in the encore phase as well, with its emphasis on enhancing family solidarity and on finding new variations on the themes of a long life.

BETTER SOCIAL CHOICES WITH AGING

The existential philosopher Albert Camus once commented, "Life is a sum of all your choices." In J. K. Rowling's book *Harry Potter and the Chamber of Secrets*, Hogwarts headmaster Aldus Dumbledore (one of the best depictions of positive aging in modern children's literature) advises young Harry: "It is our choices that show what we truly are, far more than our abilities." Maturing social intelligence contributes to better choices in every sphere of one's life, particularly in one's social life.

With age we are often more discriminating about our relationships. Research shows that older adults more readily sever superficial or unsatisfying relationships in order to spend their time with people they care about and with whom they feel comfortable and able to freely express their true selves. As one seventy-year-old woman told me, "Life is too short for me to put up with people I don't feel good being around."

Social intelligence also helps improve conflict resolution skills. Studies show that older adults use a combination of coping and negotiating strategies that lead to greater impulse control and the tendency to more effectively appraise conflict-charged situations, which results in more effective, satisfying choices of action. This is one reason that age is an asset in many people-oriented occupations such as manager, judge, politician, and diplomat.

I see this trend in many of the participants in my studies of aging. One such person, Abby Stern, told me that when she was working full-time, she never had the time or energy to deal properly with interpersonal conflicts. The pressure of deadlines (she was a magazine editor) made it hard to address issues when they arose, which resulted in pent-up emotions that exploded later on. Now, at age sixty-two and working only part-time, she can focus on conflict when it occurs and prevent molehill annoyances from becoming seething mountains of resentment. She's also more diplomatic in her words and actions, slower to respond out of hot emotion, and more likely to take a longer perspective.

"As I've grown older, my biting tongue has become less acerbic," she says.

GENDER AND SOCIAL ROLES

A number of investigators, including David Gutmann of Northwestern University, find that as men age they become more interested in social connections, whereas women, who are typically socially oriented in the realms of family and friends, tend to broaden their scope of social involvement to include the entire world and broad social justice issues. Gutmann describes men as channeling the aggressiveness of their youth into more problem-solving and peacekeeping roles. Conversely, he sees women shifting to more assertive behavior while exerting greater control in social situations. Other investigators have identified similar patterns, such as men becoming more emotionally introspective and open to relationships, and women, freed of child care responsibilities, becoming increasingly focused on broader social roles. Both these trends, researchers note, appear to promote overall health and self-esteem.

It's too simplistic to say that men and women flip roles in older age; reality is always more complicated than that. Gender roles are not so deeply divided anymore. But the changes of midlife and beyond *are* felt in a broadening and balancing of roles that for decades were shaped by dominant societal expectations. This is social intelligence (and, more broadly, developmental intelligence) in action. Men can become more whole and emotionally balanced by opening themselves to the value of social networks. Women gain by having the time and energy to take a wider view and becoming more active and assertive in societal issues.

Here's a story of one man's social growth in midlife that illustrates both social and gender-based changes.

Teaching high school chemistry had been David Conway's chief passion in life for forty years. He prepared lessons in the evenings, graded tests on the weekends, and taught summer school nearly every year. Although his dedication earned him the respect of his colleagues and students, it was one of the chief thorns in his relationship with his wife, who felt he always put work above her and their two sons.

When he was sixty-one, a new principal began to challenge David's approach to teaching. The constant battles led David to consider retiring, despite his love for his profession and his need to help support his sons, who were in graduate school. He came to me for psychotherapy because the situation was making him angry and depressed. He hated the tension and stress at school, but he was deeply worried about retiring. Not least among his concerns was how he and his wife would get along if he was around the house all the time. Their relationship was cool, strained by years of simmering resentments, and David was afraid he would simply be substituting conflict with his principal for conflict with his wife.

Still, he admitted that certain aspects of retirement were appealing.

"I told my sons about this when they were home during spring break and they said I was overdue for a change," he told me. "They said I shouldn't worry about their grad school and that I should start by retiring my nerdy wardrobe and getting ready for a new life. Our time together was better than ever. I must say I was a bit excited by the idea, but couldn't see what that new life would be like."

During our later conversations, I helped him look for a path that would respond to his seemingly incompatible desires. For example, although he was concerned about his relationship with his wife, he was open to new ideas for improving things between them. When I asked what he and his wife both liked to do together, he said "go out to eat."

"She's a connoisseur of food and likes trying out new restaurants," he said. "She'd actually love to be a restaurant reviewer for the local newspaper."

"Do you do any cooking at home?" I asked.

"No."

"Well . . . why not start?" I asked. "It's good for a couple to find something new to do together—it pumps new energy into the relationship. And, after all, don't you really deal with recipes in chemistry all the time, except you call them formulas? Why not try your hand at kitchen chemistry? It would show her you cared about that major interest in her life. Why don't you ask her? The worst she could say is 'no way.'"

The next week he said his wife looked shocked when he broached the idea, but she agreed to try it.

David also began to see that he didn't have to completely leave chemistry and teaching. He could substitute-teach or work as a tutor. That way he could continue to be socially active, transition into retirement, make some money, and gradually plan how he would begin to approach the next phase of his life.

When David filed for retirement, he said he felt liberated. He had opened a new phase in his marriage by exploring a new role, and he had also found a way to remain connected to his passion for teaching. David's maturing social intelligence led him to reexamine his roles and relationships and create a better quality of life for himself. We can also see here the Inner Push at work. I doubt that David would have made such significant changes in long-standing patterns if he hadn't felt a range of urges and drives for movement. Dissatisfactions in his life contributed to new feelings and desires to try something different—from new clothes to new cooking to a new lifestyle. Therapy helped focus and support this unfolding, but it was buoyed by developmental dynamics that transformed his social life.

The Social Portfolio

From research on healthy older adults in the past few decades, we know that mental and physical health are strongly correlated with engagement and that withdrawal and social isolation are signs of depression or some other ill. Of course, knowing the importance of social connections and being able to foster and sustain those connections are not the same thing. Making new friends and maintaining old ones can be difficult because of a lack of transportation, having poor hearing or eyesight, or having less energy to pursue sports or activities that formerly focused social gatherings.

Over the years, I have developed a number of strategies for overcoming such barriers. In their fundamentals, they echo eloquent advice that Samuel Johnson wrote in a letter to his friend Boswell in 1779 at age seventy: "If you are idle, be not solitary; if you are solitary, be not idle." This is practical social intelligence!

The approach I have developed, which I call the social portfolio, takes this advice into the twenty-first century, incorporating findings from the latest research in activity theory and brain stimulation. My idea is similar to financial advice: don't put all your eggs in one basket. Diversify your portfolio so that declines in one will be offset by gains in another. Adults need to work on a balanced social portfolio based on sound activities, mental challenge, and interpersonal relationships that they can carry into old age. I propose four portfolios, each represented in the chart below. Included are examples of activities in each of the quadrants, but these are examples only—in reality, there are as many ways to fill these boxes as there are older adults alive at any given moment!

The idea behind the social portfolio is to engage in both solitary and group activities and in activities that require high energy and mobility, such as sports, dancing, and travel, as well as low-energy/low-mobility activities, such as writing, reading, or listening to music. Four categories of equal importance are thereby created:

The Social Portfolio

	GROUP EFFORTS	INDIVIDUAL EFFORTS
HIGH MOBILITY ——— **HIGH ENERGY**	***GROUP / HIGH MOBILITY*** • Participating in an ongoing Dance or Theater Group • Traveling with a Life-Long Learning Group • Joining a Community Outreach Advocacy Group	***INDIVIDUAL / HIGH MOBILITY*** • Creating a Neighbor-hood Showcase Garden • Developing an Anno-tated Walking Tour of your Town • Doing Documentary or Nature Photography
LOW MOBILITY ——— **LOW ENERGY**	***GROUP / LOW MOBILITY*** • Forming an ongoing Best Jokes/Potluck Dinner Group • Creating Family Newspaper with Chil-dren/Grandchildren • Hosting an ongoing Book or Game Club at your home	***INDIVIDUAL / LOW MOBILITY*** • Creating the *Secret Recipes* Family Cook-book • Creating Family Tree with Dynamic Com-mentary • Creating *Ultimate* E-Mail Letters to Grandchildren

Enhancing *Individual Mastery* & *Interpersonal Growth*
Balancing *Brief* with *Enduring Relationships*

group/high mobility, group/low mobility, individual/high mobility, and individual/low mobility. You can easily create your own chart and fill in the four boxes with the activities you currently pursue. The social portfolio can help you see if your range of social engagement is lopsided—if, for instance, you need to add high-energy/high-mobility activities or if your activities are skewed too much toward solitary pursuits.

The social portfolio concept is analogous to a financial portfolio in three ways:

- It should be diversified and balanced so that the overall port-folio "performance" is steady and resilient in the face of change.
- It should provide insurance against disability or loss. If your health declines, you need interests that don't require high energy or high mobility. Similarly, if you lose a spouse or a friend, you need to have solo activities to draw on during the transition to forming new relationships.
- As with finances, your social portfolio will perform best if you start building assets early in life. But it's never too late to begin. Thus, if you have an interest in writing, you can start by taking a course, and in retirement or partial retirement, when you have more time, start writing that novel or con-tributing essays to your local newspaper. Relationships, like assets, require both time to grow and continual inputs to reach their full potential. It's great to maintain friendships made in college, but you should continue to forge new friend-ships throughout life.

Here's a quick story showing that it's never too late to break old social habits and begin forming interpersonal bonds that are so important to health. The story is about a wealthy but rather mean-spirited man who lived in London. By the mid-1800s, this businessman was well known but not well loved. One account of the time described him thus: "The cold within him froze his old features, nipped his pointed nose, shriveled his cheek, stiffened his gait, made his eyes red, his thin lips blue and spoke out shrewdly in his grating voice." The man's emotional withdrawal had evolved over the decades to the point that nobody could remember him as a happy younger man. In fact, the man was suffering from undiagnosed chronic depression, which eclipsed his real nature and long-buried social intelligence.

Fortunately, the old man was visited by a multidisciplinary team of caregivers—this more than a century before our own era of mental health outreach. This team used dream-oriented psychotherapy to help the man understand his condition—again foreshadowing by more than fifty years Freud's classic work *The Interpretation of Dreams*. They also used a form of the summing up phase life review that tapped into long-dormant inner drives to resolve conflicts and open new possibilities. With remarkable swiftness, the team's efforts paid off. The old man awakened to his social potential, staged a dramatic turnaround, and gained a vitality that enhanced his own quality of life as well as that of the people around him.

This transcendent figure is none other than Ebenezer Scrooge, the celebrated grouch of Charles Dickens's *A Christmas Carol,* written in 1843. The multidisciplinary team were the ghosts of Christmas past, present, and future. Although a work of fiction, this story reveals the real potential for change in older age, providing an example that is as valuable today as it was more than 100 years ago. No matter how late in life, no matter how severe the circumstances, when our knowledge, experience, and developmental readiness are in sync, our lives can be transformed.

Below are two stories (true ones this time!) that illustrate the course of social intelligence with aging.

ARNOLD'S STORY

At sixty-six, Arnold Rahn was forced into retirement from his job as a manager of a sporting goods store. Arnold wasn't ready for retirement and tried to find work at another store. He grew increasingly upset as he was offered one low-paying position after another. He rejected them all.

Arnold had been a low-key, easygoing man, but now, frustrated and irritable, he began to offer unsolicited advice to his two sons and criticized how they were raising their families. He also started to drink more, which made his sons increasingly reluctant to have him around his grandchildren. Arnold was reacting badly to the loss of control and self-esteem, and his maladaptive behaviors were driving his family away.

Arnold's family pulled together to brainstorm options. But Arnold adamantly opposed seeking outside help and rejected suggestions for activities not related to his previous work.

"I don't need help," he would say. "I just need to get back to my work."

By the time I became involved as a psychotherapist, the family was exasperated and Arnold was miserable. After listening to all sides of the situation, it occurred to me that Arnold might like teaching business or management courses at the community college level. His sons were skeptical. They thought Arnold would dismiss it out of hand; they also worried that if he applied for such a position and was rejected, it would make things worse. I offered to "test the waters" with some programs I knew about from prior experience. I found an instructor who was happy to invite Arnold to lecture in his class so the students could learn from his real-world business experience, and with a bit of coaxing, Arnold agreed to give it a try.

The class he taught was a huge success, and he was invited back repeatedly. It turned out that the intensely social experience of teaching was just what Arnold needed. His mood and attitude improved, which, in turn, improved his relationship with his family. Arnold's native social intelligence had been hijacked by the negative feelings associated with his forced retirement. Fortunately, his downward spiral was arrested before great harm was done to his interpersonal relationships. He had the gumption to seize an opportunity and then use his expertise and social skills to full advantage in a new environment.

AGNES'S STORY

Agnes Rafferty was the oldest of seven children. When her mother died young from pneumonia, Agnes took on many maternal responsibilities, helping her father raise the youngest children. Despite the workload, Agnes finished high school. She delayed going to college for several years to continue helping her father, but eventually she finished college and went on to get a master's degree in English literature. Like her mother, Agnes became a schoolteacher.

She taught English literature with flair and charisma and won many Best Teacher awards over the course of a long career. But her personal life was difficult. Her husband, with whom she had two sons, died from a heart attack when the children were young, leaving her, again, to shoulder dual responsibilities. Once again, her fierce determination and independent spirit prevailed and she raised her sons while continuing to teach.

By the time she was forced to retire at sixty-five, her sons had graduated from college, married, and moved to California. She was settled in her life, though hardly ready to slow down.

"I am not going to stop doing things just because I'm sixty-five and have to leave the school," she said.

Encouraged by two former students who were now local librarians, Agnes launched a series of readings and talks about literature. The program, "Afternoons with Agnes: Voyages in Twentieth-Century Literature," gave Agnes a platform to do what she loved: read great works with the dramatic characterizations of a stage actress.

Then, in her mid-seventies, Agnes had a stroke. Her crisp, articulate voice became slurred, and she needed to use a walker. Her sons urged her to move closer to them, but Agnes declined; she felt rooted in her town and social networks. She clung to her independence even though she was clearly suffering. She stopped doing her literature programs because she didn't like the idea of reading sitting down and was embarrassed by her impaired speech.

She began to withdraw from life. Her appetite and weight declined.

When her family asked me to intervene, I saw in her a spark of life that flickered through her depression. Her apartment was filled with books, and as I browsed, she asked what my favorite book had been in high school. I said Hemingway's *Old Man and the Sea*. She said that was always one of her favorites too. And then she added that she now felt like she had that monstrous sea creature on her back and in her mouth.

"You have such a great way with words," I said. "Have you done much writing?"

"My tongue works better than my pen," she replied. "I love literature and know so much more now than when I started to teach," she continued. "I loved sharing it, but I can't any more."

I could see she was caught up in a negative swirl of thoughts that sapped her motivation to try something new. I left that day with an idea. I contacted the librarians and broached an idea for a weekly bulletin from Agnes called something like "Advice from Agnes—for Voyages in Twentieth-Century Literature." Each flyer would recommend a work of twentieth-century literature and would include her comments about what was special or engaging about the work as well as a brief excerpt.

The librarians loved the idea and proposed it to Agnes. To help her overcome her insecurity about writing, I suggested that since she was so good at oral presentations, she tape herself and then use the tape as the basis for the flyer. She agreed to give it a try.

"Advice from Agnes" soon became a regular fixture of the library system. Although she no longer stood in front of groups of people, Agnes exchanged phone calls and letters with her readers. She began to reweave the fabric of her social sphere and soon regained her spirited, independent identity and spunk.

SUMMARY

Social intelligence, memory, and wisdom are closely related fruits that age alone can ripen. The aging brain has greater potential than most people think, and development never stops. Our capacity for social involvement and interpersonal relations remains as strong as ever in later years and is a vital wellspring of both physical and mental health. The final two chapters will bring together these positive aspects of aging and apply them to two fundamentally important realms: retirement and creativity. As with practically every other component of aging, both areas are rife with myth, misunderstanding, and needlessly negative expectations.

7

Reinventing Retirement

I never remember feeling tired by work, though
idleness exhausts me completely.

—Arthur Conan Doyle

JUNANN HOLMES, SEVENTY-ONE, awoke to rustling sounds at the
foot of her bed. Lifting her head to see more clearly, she was star-
tled by two pairs of eyes peering intently back at her. It took her
a moment to remember that she wasn't in her Washington, D.C.,
bedroom. In fact, she was in Borneo, and the eyes belonged to
two red-haired apes—a mother orangutan and a baby, held in her
arms.

JunAnn was volunteering for the Orangutan Foundation International, which has a research program on the endangered
animals. The mother orangutan had smelled the fruit juice on the
table beside the bed. She had pushed open the door of JunAnn's

hut and was now leaving, the juice container in her hand. JunAnn chuckled and went back to sleep. Another evening in Borneo.

JunAnn had "retired" years earlier, which is why I'm telling her story here. Retirement isn't what it used to be. In fact, the word "retire" hardly applies to what I am seeing in the lives of older adults today. "Retire" connotes a disengagement from the world, a withdrawal of activity, a "decommissioning" of a human being into a mothballed state of repose.

JunAnn, like hundreds of thousands of other older adults, rejected "retirement" and all it stands for. I met her four years before her interrupted sleep that night in Borneo in the course of conducting an extensive study titled Retirement in the Twenty-first Century. JunAnn grew up in a series of foster families because her own family couldn't care for her during the Depression. Despite this fractured upbringing, she finished high school and college, majoring in special education. She never taught, however. Instead, she worked at a variety of clerical jobs before joining an airline as a ticket agent and customer service representative. Whenever she could, JunAnn volunteered to tutor children with special needs. She never married and had no children, but children always liked being around her, especially when she read to them in her warm, appealing voice.

In her fifties she started volunteering at Washington's National Zoo. She took a particular interest in the apes and elephants—and they seemed to reciprocate her curiosity. At one point she heard a lecture about orangutans by professor Biruté Mary Galdikas and was enthralled by her work. Heeding an internal sense of "if not now, when?" JunAnn joined the Orangutan Foundation. At first she just attended meetings and read material, but she then decided she wanted more direct involvement. She volunteered to assist Galdikas. Over the next dozen years, JunAnn traveled to Borneo many times to better understand the behavior of orangutans, and especially the connection between mother orangutans and their offspring. JunAnn became deeply involved in caring for orphaned orangutans, and she says the entire experience was one of the most fulfilling of her life.

Like the apes in the zoo, the wild orangutans were very comfortable with her. The very young ones, who needed to be held like young human babies, would cling to her. Older orangutans would leap on her from the trees when they spotted her passing by. Her own experience in foster families had endowed her with deep empathy for these fellow primates separated from their mothers.

For JunAnn, "retirement" was a beginning, not an end. She is one of those people for whom the liberation phase solidifies a sense of personal identity, clarifies life goals, and empowers one to make a difference in the world. Like JunAnn, people all over the world are literally reinventing retirement.

Before going any further, however, I want to make something clear. By featuring JunAnn and focusing, as I will, on people who take an active role in "retirement," I am not saying that everybody should be this active. Some people, particularly those who have labored hard for decades, may not want to be active, at least for a while. Some sociologists and gerontologists, for example, have described a "busy ethic" in which older people feel compelled to say they are keeping busy as a way of defending their leisure time. David J. Ekerdt, director of the gerontology center at the

University of Kansas, coined the term "busy ethic" in a 1986 paper. He warns that, like everything else, high-energy activity can be taken to an extreme, with a subsequent devaluation of quieter forms of engagement.

Indeed, some people simply prefer to be inactive.

"I hate people who say, 'Now I'm going to college and I'm going to go bungee jumping and have sex till I'm eighty,'" says Virginia Ironside, an advice columnist for the London *Independent* who is writing a book about the pleasures of doing little in old age. "Now is the time to wind down. I've bungee jumped till I'm blue in the face, metaphorically."

I can appreciate Ms. Ironside's perspective—and I admire the pugnacity with which she expresses her views. To her and others like her, I say "Go for it!" If you want to simply relax and do nothing in particular after you quit full-time work, by all means, do that—you've earned it! I can't help noticing, however, that despite Ms. Ironside's avowed desire to "wind down," she has just finished writing a book (at age sixty) called *No, I Don't Want to Join a Book Club*, in which she elaborates on her rejection of the so-called super-retirement. This is not what one might expect from a woman who says that all she wants to do is "piddle around."

Still, I think her point is well taken. No one should feel pressured to be more active than they want to be. In my experience, however, I don't find many people like this. On the contrary, time and again I meet people who want to be *more* active, *more* engaged, and *more* stimulated. No one is pushing them to this—it is a natural outgrowth of inner desires we all have for learning, social relationships, a sense of meaning, and giving something back to society.

Reinventing Retirement

The face of retirement is changing because people are following their own instincts to do more than sit around and play bridge with their peers (though bridge is good for the mind). For decades, the age of retirement slowly dropped during the twentieth century; then in the 1980s the trend reversed, and since then, retirement age has been rising. The participation of older men in the workforce has remained stable, whereas older women's participation rates have begun rising dramatically. In recent years, many public policies and private institutions that encourage early retirement have been modified. Mandatory retirement was outlawed in most jobs. Social Security is no longer growing more generous, and coverage under company pension plans is no longer rising. In addition, both Social Security and private pensions have become more "age neutral," meaning that they provide either weaker incentives or no incentives to retire at particular ages, such as sixty-two or sixty-five.

Between 2002 and 2012, the number of workers aged fifty-five and older is expected to grow by 49 percent while the number of workers under fifty-five will grow by only 5 percent. The United States has one of the highest labor force participation rates for persons aged sixty-five and older in the developed world, surpassed in 1999–2000 only by Japan, Iceland, and Portugal.

In other words, the dividing line between "career" and "retirement" is not only moving into higher ages, it is also becoming more blurry, with more people opting for phased retirement options that allow them to work part-time while still receiving some benefits. Some people never retire in the classic sense; they continue writing or teaching or coaching or performing until the end of their lives. And not because they have to, but because they *want* to.

Retirement is also being reinvented in social and psychological ways. Despite the stubborn retention of the notion that older people are "over-the-hill," it's becoming increasingly clear that the second half of life can be more rewarding, stimulating, enjoyable, and rich than the first half.

MY STUDY OF TWENTY-FIRST-CENTURY RETIREMENT

As I write these words, the in-depth study of modern retirement I began in 2000 is in its fifth year. Today, more than 100 adults over age sixty have participated. All were either retired, partially retired, or within a year of retirement when they enrolled in the study. The participants fairly represent the larger population of older adults in terms of gender, race, and income level.

My goal has been to get an in-depth view of each participant: what they value, how they view themselves and retirement, what they are doing with their lives, and how they are responding to the developmental changes of older age. But I don't simply want snapshots—I want to see if things change over time. That requires multiple face-to-face interviews, all of which I conduct myself. Although this is a lot of work, it is also incredibly rewarding and enlightening. Not only do I enjoy the process of getting to know the study participants, but I also constantly learn new things from them. Indeed, I am in a privileged position as a research interviewer. I am able to ask questions that friends or relatives might shy away from, such as "What are some of your fears about growing older?" or "How do you feel about the loss of your spouse?"

The interviews combine a set of questions that I ask everyone as well as many open-ended questions that allow people to "fill in the blanks" however they want. The value of open-ended questions is that you never know what insights you may uncover. The following anecdote illustrates my point.

Mary Leahy was rattled when she arrived at my office late. She told me about the delays that morning on the subway. Instead of taking the lead to "get down to business" with questions for the study, I let her talk. She mentioned that as she was waiting for a train, she picked up a pamphlet called the "Metro Pocket Guide," which listed points of interest by each Metro stop. She started telling me about how she never knew there was so much to see and do along the subway route she had traveled many times. Then she paused. "Am I boring you?" she asked.

"Not at all," I said. "In fact, you've just taught me something that I think more people should know about."

I had not heard of the Metro guides and it struck me as a good idea with potential value to older adults (and younger adults too) who live anywhere there is public transportation of any kind—subways, buses, or trains.

"Obviously there are a lot of people like you who use public transit," I said. "This guide you describe is a great idea. I think every locality with public transportation should produce something like this—and I'll spread this idea when I talk to people around the country. Thanks for sharing!"

This is what can happen in personal interviews that provide ample opportunity for unstructured discussion. To date, I have conducted more than 1,000 hours of interviews, which provide a unique and valuable view of retirement in twenty-first-century America.

SOME PRELIMINARY RESULTS

The picture of "retirement" emerging from my study vividly contradicts the myths and prevailing assumptions so rampant in our culture. "Retirement" simply doesn't describe the reality of most of the people I have interviewed. These people are anything but "over-the-hill." Indeed, most (though certainly not all) are climbing *new* hills, not coasting down previously climbed hills. They are charged with a new sense of personal adventure, which I see as one product of the Inner Push I've discussed in previous chapters. People in each of the four human potential phases of later life are finding unique ways to harness the potential latent in changes such as ending full-time work, seeing children leave the house, or even losing one's spouse or close friends to death or illness.

As a culture we haven't come up with a good word or phrase to replace "retirement." "The Golden Years," while not quite right and a bit sappy, at least has positive connotations of worth, value, and reward. Whatever term eventually replaces "retirement"—and I am sure such a linguistic evolution will occur sooner or later—it will reflect the realities that I have found in my study. (Mine is not the only study of retirement, of course, and I should add here that my findings are being corroborated by other studies.) In my study, for instance, 37 percent of participants in the "retirement period" were only partially retired. This is similar to the findings of Phyllis Moen's earlier Cornell Retirement and Well-Being Study, which surveyed retirees between the ages of fifty and seventy-two (as compared with sixty to ninety in my study). In the slightly younger Cornell group, 44 percent were partially retired. These partially retired people continued to work part-time, either in new business activities or in seasonal work. Strikingly, more than half of those I interviewed—women and men alike, up to the age of seventy-five—said they would like to work at least part-time if the right job were available. This means, of course, that the other half are *not* interested in returning to work. But still, it seems clear that many older adults are unsatisfied by unemployment in retirement.

The one aspect of the traditional notion of "retirement" that has remained valid is the general availability of more personal time. People in my study say that not having to work full-time or not having children to look after on a daily basis allows them time to take stock of their lives in a way they've never experienced before. With less stress and pressure, many older adults explore new options and experiment with new activities. Many comment on how empowering it feels to be more in charge of their own time, to "be their own boss."

Surprising and valuable insights have emerged from the study. The lessons, for the most part, are only now beginning to be widely recognized and incorporated into retirement plans made by individuals, business leaders, communities, and policy makers. We still have a long way to go in this country to capitalize on the potential of the second half of life, to reduce barriers to personal self-fulfillment, to ease the social and physical isolation that so often plagues older adults, and to shift from a negative view of aging to one based on the positive realities I see all the time in my research subjects.

Lesson 1: The Need for Planning

Most people facing retirement—or even partial retirement—have not done any planning for this major life transition. I'm not talking about financial planning—which is, thankfully, something more people are taking seriously from an early age. I'm talking about planning for how you will be socially engaged, how you will spend your time, what larger goals you want to pursue, and how you can take full advantage of the extra time available in this phase of life. This finding, too, is similar to that in the Cornell Retirement and Well-Being Study, which found that when planning did occur, it was typically in the financial area and not in the personal time arena. Less than 10 percent of my study participants had done any preparation beyond financial planning. Yet almost everyone I've interviewed said that they would have benefited greatly from some education about how to develop a social portfolio, as discussed in the previous chapter.

A logical place to receive preretirement advice and education would be the workplace, but only a handful of the top 100 of the Fortune 500 companies have anything like a complete program for retirees. The lack of planning and preparation for retirement undermines people's opportunity to broaden their horizons with novel recreational activities, educational enrichment, and civic engagement. Companies spend vast amounts of money to improve and preserve the health of their employees by providing them with comprehensive health care plans. Helping employees deal more effectively with the transition from full-time work to retirement is consistent with that mission.

When I say "retirement plan," I don't necessarily mean that you need a road map of specific steps, activities, and goals for the rest of your life. Some people may, indeed, be able to plot out their options in great detail, but a "plan" for others may be loose and open-ended. Some people want to experience their new transition first, and *then* chart new paths. For them, exploration is their "plan." After all, as the saying goes, "We don't know what we don't know." Retiring can be like going to college, which is also an open-ended process. In college we are exploring options, exposing ourselves to new ideas and new adventures. Many colleges recognize the value of casting a wide net, of dabbling in courses, and thus they don't require commitment to a major until the sophomore or junior year. You can enter retirement the same way—exploring, dabbling, trying this or that in order to see what appeals to you. After a period of active exploration, you may (or may not) come up with a focused plan for your "retirement."

Lesson 2: The Need for Community Infrastructure
Related to the problem of poor planning is the enormous gap in what communities provide to help people match their interests and skills to local needs. I see many people with extraordinary expertise who don't use their skills because they don't know where to offer their help. An analogue of traditional "help wanted" ads does not exist for volunteer opportunities. I think that community leaders and policy makers have an opportunity and a responsibility to help retirees in this matching process. The result would be good for everyone. Perhaps as older adults become more familiar with the

Internet, it will be easier to create community "volunteers wanted" Web sites that could serve this function. (See appendix 2 for information about national organizations dedicated to older adults, retirees, and volunteers.)

Lesson 3: The Value of a Balanced Social Portfolio

I've been struck by the imbalance in the way many older adults socialize. Some spend the bulk of their time alone—they could benefit from group activities. Others are almost exclusively active—always going, going, going; they would benefit from cultivating some nonenergetic activities. Relatively few people had a well-balanced "social portfolio" of the type I described in the previous chapter. One who did was a seventy-nine-year-old woman who had joined a bridge club to challenge her mind and stay socially engaged (a sedentary activity) while also signing up for "lifelong learning" courses that involved travel (an energetic activity).

Policy makers and the leaders of adult-oriented organizations should think hard about this finding because it highlights a need for education, guidance, and support. The goal of "getting older adults involved" is often construed too narrowly. It's not enough that a series of lectures be presented, for instance. Opportunities for energetic activities need to be offered as well. And, more to the point, people who tend to favor one type of activity over another should be encouraged to try a different kind of experience. The avid bird-watcher who is always taking walks might be encouraged to add less energetic activities to his or her schedule, for example. Having a mix of private vs. group and energetic vs. quiet activities more broadly stimulates your brain and body than focusing on one type of activity or another.

Lesson 4: More Engagement over Time

During the first three years of the study, I tracked the number of activities in which each participant was engaged. To my surprise— and in complete contradiction to the "disengagement" theories I reviewed in the previous chapter—I found that the trend among these subjects was for *increasing* levels of activity over time. In fact, the third year was when most subjects reported their greatest social involvement. And many reported that the third year was their "best year" of the three I studied.

This trend may be due in part to the fact that these people were in the study. By encouraging them to review their experiences and consider options, I may have stimulated them to become more active than they would have been otherwise. This kind of study side effect can't be discounted, but neither does it explain every-thing. After all, my interviews accounted for only a tiny fraction of the participants' lives, and such a relatively small intervention is highly unlikely to account for the overall rise in activity levels I observed. (To the extent that our sessions did influence the nature and magnitude of their community involvement, I see this as evi-dence for the beneficial effects of providing a formal infrastructure of programs designed to assist older persons in their deliberations about volunteer or work options.)

Rather than being stimulated by the study itself, I think the participants were responding to their Inner Push to seek more social connections and mental stimulation. I'm reminded of one married couple I interviewed. Both in their mid-sixties, they sought out unusual activities in their community. They discovered a pro-gram offered through their local police department in which residents were invited to attend workshops designed to give them a hands-on understanding of the responsibilities, challenges, and experiences of police work. One of their sessions included spending a day in a patrol car to witness a day in the life of a police officer. This couple thrived on these kinds of activities. They enjoyed meeting new people, seeing things from new perspectives, and challenging themselves mentally and physically.

Lesson 5: The Value of Long-Duration Activities
Time and again I have found evidence that the duration of an activity is more important than the nature of the activity itself. In other words, a book club that meets on a regular basis over the course of months or years contributes a great deal more to a person's well-being than the same number of one-shot activities, such as movies, lectures, or outings. Long-duration or repeating activities are better for establishing new relationships and making new friends. Hence, a balanced social portfolio includes brief nonenergetic activities and enduring programs that involve others. This lesson has not yet been learned by many older individuals and those responsible for helping create programs or activities for them.

The value of repeating activities is illustrated by one of my study participants, a devoted opera buff. Although he often attended the opera, he found that he didn't make new friends that way. So he formed an opera video and dinner club. He hosted evenings at his house where opera lovers could gather to watch an opera on television, with dinner and discussion between acts. Opera thus became a important social focus in his life.

Lesson 6: The Difficulty of Forming Close Friendships
Related to the previous finding is another trend: the difficulty of
finding new friends, particularly new close friends—people with
whom you feel comfortable confiding personal issues, problems,
fears, or joys. In psychological parlance, we call such friendships
"confiding relationships." Research shows that having at least
one—and preferably more than one—confiding relationship is a
key ingredient of mental health and resilience.

Many older adults tell me they don't have opportunities for
making new friends. Sometimes this is because they simply aren't
looking very hard. Many communities, even small ones, have
organizations, programs, centers, or other types of gathering places
that are great venues for meeting people if you just look for them.
But it's also true that it can be difficult to find such places or programs because of the lack of centralized, well-publicized listings of
such opportunities. Again, this is a lesson as much for older adults
as for the people in charge of senior centers, local area agencies on
aging, and other organizations dedicated to enriching the lives of
older adults.

To older adults reading this book I offer the following hopeful
observation: Making new friends and establishing new social connections can be a self-reinforcing phenomenon. The more people
you meet, the more people you meet! Getting involved in just one
group can open you up to news about other gatherings. I frequently
see this kind of social cross-fertilization. One woman I know who
retired from her job as a crossing guard in her mid-sixties, slowly
expanded her social portfolio. She started by hooking up with a
mall-walk exercise group—people who met regularly to do laps

around a shopping mall as part of a program that offered them a free movie ticket each time they met. Then she and her husband heard of a recreational vehicle club that made outings to scenic areas for long-weekend group camping adventures. Now she and her husband try to travel this way once a month from April through October. The friends she has made in these groups get together weekly at one another's homes for dinner, games, and cards. Of course, she isn't on the go all the time; she and her husband have plenty of time for R&R, but they are examples of people who have developed a very nicely balanced social portfolio by following up on leads generated by an initial foray into a group activity.

Lesson 7: The Importance of Giving Back

The most thought-provoking question I ask about retirement is "What gives you a sense of meaning or purpose in life?" The nearly universal response is "making a contribution and helping others." I hear this response from people of all income levels, all races, and all cultural backgrounds. It signifies a fundamentally charitable and noble human impulse: to contribute to the greater good. As we learned in the chapter about the summing up phase, the impulse to give back becomes especially acute in later life as people's perspectives about their own mortality shift and as their values change as a result of confronting the challenges of aging. Not everyone acts on this desire, often because they don't know where to offer their help. Communities need to make it easier for people to find outlets for their charitable impulses. But the desires exist and constitute a tremendous societal resource.

Participants in my study who found meaningful volunteer experiences and other ways to "give back" were those most satisfied with their retirement. The group most at risk of dissatisfaction in retirement were those who had careers in which they had been making a difference in the world but who found it difficult to achieve a similar sense of fulfillment in retirement. This group would have benefited from a preretirement planning program—though, as I've noted, these are usually lacking in most communities and large corporations.

An example of the positive effects of planning can be seen in the story of a retired schoolteacher in one of my studies. She loved teaching and had been involved for years as a volunteer in her national professional society. When she retired, she became an officer in that organization. "I might not have retired for at least another five years had I not been able to assume this new role," she said. "It's turned out to be a great move and very rewarding."

A few communities are starting to explore ways to connect talented retirees with community volunteer experiences. For example, I've worked with an innovative group in Montgomery County, Maryland, called Senior Leadership Montgomery. The group arranges both learning programs and "community action projects" aimed at meeting specific needs, such as help in schools, libraries, the court system, or recreation programs. The program casts a wide net: It's open to anyone older than fifty-five, retired or not, who is "ready to share their wisdom and talent with like-minded individuals who want to make a difference." (For more information on this group, see appendix 2.)

Of course, some people find ways to create their own volunteer opportunities. Charles Vetter, for example, retired from the U.S. Information Agency, where he was an expert on the former Soviet Union. He cared deeply about Russian history, politics, and culture and wanted to share that knowledge with others in the hopes that greater familiarity with this old adversary would improve relations between the two countries. He came up with a fascinating idea: He invented a character named Alexandr Petrovich Surov, a retired Soviet government official. He practiced a Russian accent and created a detailed "biography" of his character on which to draw in appearances before audiences. He would walk onstage and explain in his "Russian accent" that he was substituting for Dr. Charles Vetter, who had to cancel at the last minute. He would then regale the audience with fascinating insider "anecdotes," summaries of current events from the "Russian" perspective, and stories from the Cold War. He kept abreast of Russian culture by constant reading and attending lectures at area think tanks, such as the Woodrow Wilson International Center for Scholars. He would only reveal his true identity at the end of the lecture, which invariably surprised and captivated the audience. In his eightieth year, Vetter made over 125 appearances, either as Mr. Surov or as himself giving presentations on other topics to student and community groups—an amazing example of creating a new social network in "retirement" as well as a fine illustration of how one person channeled his innate urge to give something back to society in his later years.

Lesson 8: The Importance of Lifelong Learning

Another message emerging loud and clear from my data is the high value most retirement-age individuals place on lifelong learning, in forms as varied as formal classes, lectures, travel-related education programs, workshops, clubs, and other groups. I found more desire for such experiences than actual participation, mostly because of financial limitations. Many older adults want to exercise their minds but are constrained because education usually costs money. This is another important lesson for society. Communities should plan more low-cost and no-cost educational opportunities for older adults. After all, older persons continue to pay school taxes, but they do not reap any educational benefits from this support.

This study and others show that educational activities that contribute to a sense of mastery promote health and independence in later life. Hence, creating educational opportunities for everyone is good in and of itself as well as pragmatic because it will reduce the risk factors that contribute to the institutionalization of older adults.

One participant in the retirement study found ways to satisfy her appetite for education despite a limited income. This sixty-two-year-old retired secretary took advantage of living in Washington, D.C., by asking foreign embassies about no-cost public cultural events. She was never at a loss for something to do! But you don't need to live in Washington to find such activities. Local colleges, museums, libraries, and other educational centers often have no-cost or low-cost events and activities—although these are seldom publicized as widely as they could be.

Communities can also do more. For example, a program could encourage people with season tickets to cultural or sports events to donate unused tickets to senior centers. Area theaters could offer senior citizens free admission to dress rehearsals. Older persons could also be invited to volunteer as ushers in exchange for free access to performances. Printed guides to local cultural, historic, scenic, and recreational interests could also be created, similar to the Washington, D.C., Metro Pocket Guide mentioned earlier.

THE RETIREMENT READINESS QUOTIENT

I give the people in my study the following questionnaire to help them determine their Retirement Readiness Quotient on a scale of 0 (not at all ready) to 12 (as ready as you can be). Take some time to think about the questions and the accompanying explanations of their possible significance. If you have not given much thought to any of these questions, or if you do not have many good answers for them, you are probably not well prepared for retirement.

1. Why are you thinking about retirement now?
   (One point if when you write it down and then read it out loud to yourself, it makes sense, or if someone who is reliable and knows you considers your answer good and clear. Zero points if your reasoning seems fuzzy or you are simply uncertain.)
   Significance: Your reasons for retirement should be sound and not impulsive or the result of inadequate planning.

2. Do you really want to retire?

   (One point if yes, zero if no.)

   Significance: This seemingly simple question is an excellent predictor of success in the transition to retirement. It asks you to consider your deepest desires and motivations, not just what you "think" you ought to do or what other people expect you to do.

3. What do your family and friends say about you retiring?

   (One point if they think you are doing the right thing.)

   Significance: Feedback from those who know you well can be invaluable when you're contemplating retirement. Do they think it's a good decision? Do they think you have thought it out well and prepared sufficiently for it?

4. Have you considered whether you want a complete or partial retirement? Have you considered part-time or temporary work, or even a less-than-full-time small business venture? (The emphasis here is on *consideration*.)

   (One point if yes, you've considered the options, even if you choose to retire completely and not go the partial route.)

   Significance: If you are not entirely sure about retiring or are concerned about finances, then phased, or partial retirement is an important option to consider.

5. Are your finances sufficient to carry you through your retirement years while continuing to enjoy your current lifestyle?

(One point if yes to both parts of the question; zero if no to either part.)

Significance: If you answered no, you clearly have further financial planning to do.

6. Have you attended a retirement preparation program or seminar focused on financial planning?

(One point if yes, zero if no.)

Significance: Such programs can help you plan spending, predict future income, and anticipate future needs. A bewildering number of options exist, and getting some objective advice is invaluable.

7. What gives you a sense of meaning and purpose in life?

(One point if when you write it down and then read it out loud to yourself you feel you have adequately identified what gives you a sense of meaning and purpose in life. Zero points if your reasoning seems fuzzy or you are simply uncertain.)

Significance: A lack of clarity about your core values and what aspects of life hold meaning for you is often associated with a less fulfilling retirement.

8. What specific types of activities and experiences are important and fulfilling for you?

(One point if when you write it down and then read it out loud to yourself, your description of how your plans relate to what is important to you makes sense, or one point if someone who is reliable and knows you considers your answer good and clear.)

Significance: This is a more specific version of question 7. Your answers here provide a window on how well you really know your mind and how well you have planned how to accomplish what is important to you.

9. Have you attended a retirement preparation program or seminar focused on social planning (e.g., community activities and interpersonal endeavors)?
   (One point if yes, zero if no.)
   Significance: Prospective retirees often fail to adequately plan how they will actually spend their time in retirement. Floundering in these areas, without adequate preparation, can lead to frustration and a disappointing retirement life.

10. Have you developed any outside interests, hobbies, volunteer activities, or areas of new learning?
    (One point if yes, zero if no.)
    Significance: Developing new interests can improve the quality of retirement life, and engaging in challenging new endeavors can present new opportunities for personal mastery and empowerment that are associated with positive health outcomes.

11. Have you planned new activities that would allow you to interact with people on a regular basis and that offer chances to form new friendships?
    (One point if yes, zero if no.)
    Significance: Making new friends is often more difficult in

*retirement, and loneliness is associated with a host of mental
and physical ills.*

12. During retirement, will making only a modest contribution in
volunteer activities be sufficient for you?
(One point if yes, zero if no.)
*Significance: People who have had satisfying and personally
meaningful careers can find the transition to retirement difficult if they do not plan for other ways to make a
difference. Such people might consider a phased retirement
so they can continue with fulfilling work while starting
their retirement.*

Scoring

12 points: You're in position for a great retirement.
10–11 points: Your retirement will likely be highly satisfying.
8–9 points: Your retirement could have problems that are
                likely fixable.
6–7 points: You could be challenged by ambivalent feelings
                about retirement, requiring a solid effort to bring
                your situation up a notch.
3–5 points: You are potentially in the trouble zone where
                your retirement might not work well unless you
                make a major effort to get it on track.
0–2 points: You are in jeopardy of having an unfulfilling
                retirement, requiring an all-out effort to improve
                your retirement prospects.

CHALLENGING YOUR MIND AND
IMPROVING HEALTH IN RETIREMENT

When people hear about my retirement study, they often ask what they can do to improve their mental and physical health in this phase of life. Of course, there are many things you can do, but since we human beings seem to have a predilection for the number ten when it comes to such things as lists, commandments, and ratings, I will present the Top Ten ways to stay mentally and physically fit in later life. I can't claim that I have brought these suggestions down from the mountain, but they *are* based on a mountain of data, including the latest findings from neuroscience.

1. *Play games and do puzzles.* Word games such as crossword puzzles or Scrabble are particularly useful, although any type of game that requires mental manipulation or recall of facts is helpful. Studies show that you can continue to increase your vocabulary throughout life. More generally, games such as bridge and other games that require memory and strategy facilitate a quickness of mind and create new synapses.

2. *Plan a dinner and book or video discussion group.* Provocative discussion and food for thought in a stimulating and entertaining social atmosphere is as good for the mind as it is for the palate.

3. *Travel to someplace new—local or distant.* Think of a new place you'd like to visit and go there, either alone or with friends or family. It can be as close as a new museum exhibit in town or

as far away as the Orient. Consider recording your experience somehow—in writing, audio recording, sketches, photographs, or video.

4. *Enroll in a course about an unfamiliar subject.* Lifelong learning is lifelong growth and development. Today's increasing intergenerational mix in continuing education classrooms provides opportunities for gaining interesting new knowledge and new relationships. You can also combine learning and travel through programs like Elderhostel.

5. *Explore the hobby or crafts section at a bookstore for new ideas.* Even if you're not sure what you might be interested in, browsing through books on hobbies and crafts may ignite a new curiosity or remind you of a long-standing interest that you never had time to pursue.

6. *Volunteer.* Volunteering is a way of sharing special skills or learning new ones while interacting with people and providing community service. It can be an avenue for experimenting with new approaches and for working with all age groups. Even among people in their early eighties, more than one-fourth still volunteer. Civic engagement has both high personal and social value.

Don't forget that these days you can also volunteer via the Internet. There are Web sites for every conceivable interest, and many of them have chat rooms in which more experienced members answer questions or help solve problems faced by less experienced people. If you have a particular expertise, you may find yourself in high demand. An example of an online volunteer activity is Wikipedia (www.wikipedia.org), an encyclopedia that anyone can add to or edit. Thousands of ordinary people who happen to have a particular interest in or knowledge about a specific topic volunteer to write articles, correct mistakes, edit grammar, and otherwise improve the quality of this phenomenal new resource.

7. *Consider new part-time work.* Many "retired" people continue to work on a part-time or temporary basis, either for the money, the social stimulation, or both. Attitudes toward older workers are improving, especially in our expanding service-oriented society, where the experience and conscientiousness of seasoned workers pays off. Check with your local librarian for job leads or explore some of the Web sites listed in the appendixes of this book.

8. *Correspond with family and friends.* Setting a regular schedule for writing letters or e-mails to family members or friends can not only strengthen your web of relationship connections, it's also excellent exercise for your brain. In this digital age, remember that getting a "real" letter in the mail can be special to the recipient. But with e-mail you may be more likely to get a response from busy children or (especially) grandchildren than to a "real" letter.

9. *Develop a dream journal.* Dreams and daydreams are the most accessible ports to our inner creativity. Write them down or draw them. They may open your eyes to inner thoughts and desires and help you tap your creative potential. Keep your journal by your bed because dreams fade quickly upon waking. You can also use audio tapes. You don't have to record every dream, or even everything about a dream (often dreams

are so full of details that describing everything in words can be taxing). Just record what seems especially interesting, bizarre, or personally meaningful. You don't have to interpret your dreams either. Often dreams are just dreams—they don't "mean" anything in particular. They *may* reveal unconscious urges, conflicts, or emotions, but then again, they may simply be your brain's way of cutting loose and processing all of the images, emotions, and thoughts of the day in the illogical, random, and surreal form of dreams.

As an example, recall from chapter 3 the story of James Dunton, the computer engineer who in midlife was unsatisfied with his career. He had a series of disturbing dreams, in which electrical fires and other calamities destroyed computer-filled offices. He felt the dreams were telling him that it was dangerous for his mental health to continue the work and lifestyle he was in. He decided to pursue teaching and realized in retrospect that it was the right move at the right time in his life.

10. *Write or record your memoirs, autobiography, or family history.* Autobiographies are not just for the famous. Developing a genealogy, family history, or memoir provides a valuable gift for your family or friends. It can also launch you on a new journey of personal exploration and discovery, getting you in touch with historical and psychological roots. Consider "talking" about your past to a tape or digital recorder—this approach has the advantage of preserving your voice as well as your words. If transcribing the tapes seems a task beyond your time or abilities, ask for help. You may be surprised at how willing friends or family members are to help preserve your recollections.

Genealogy can also be a surprisingly social activity. The Internet is alive with people discovering long-lost (or never known) family connections. The resources available today for tracing your ancestry are astounding—census records, cemetery listings, ship passenger logs, birth and death certificates, and much more are available at the click of a mouse. It can be thrilling to link up with "cousins" you didn't know you had and fascinating to learn more about your own heritage.

This Top Ten list is really just a start. It contains some specific suggestions that I have found are particularly good at stimulating the mind, creativity, and social interactions. But the potential list is endless. The important thing is to start. As Aristotle said, "The beginning is half the whole."

Summing Up Retirement

The crisp dividing line between "career" and "retirement" is becoming increasingly blurred with such ideas as phased retirement and continuing part-time or temporary work by older adults. And the lifestyles of those who no longer work full-time are richer, more active, and more engaged than in previous eras. More and more people are realizing that this time of life, far from being the downside of the "hill," is full of potential for improving life, expanding skills, and fulfilling dreams and expectations.

Unfortunately, many people remain unprepared, financially and psychologically, for the transition of "retirement." Virtually all of my study participants say they would have taken advantage of a good retirement preparation program if it were offered, but this is still a vast unmet need. We need more programs like the retirement planning workshops offered by the North Carolina Center for Creative Retirement, directed by Ronald Manheimer (see appendix 2).

Most communities lack the societal infrastructure to help prospective retirees plan for and navigate retirement or partial retirement. Few volunteer placement or job placement programs exist for older workers. Although volunteer opportunities sometimes abound, no road map or organization matches opportunities with people who have valuable skills and a desire to help. The result is a vast underutilization of the great national resource that older persons represent.

If you or a loved one is retired or is considering retirement, remember these key points:

- Look for activities that involve repeated interactions over time. Some people are very busy in retirement but still feel alone because they fill their time with transient or non-interpersonal activities that don't give them opportunities to meet new friends.
- Try over time to develop a social portfolio balanced between energetic and nonenergetic activities and solo and social activities.
- Find ways to challenge your mind and body.
- Do all you can to remain physically fit. Keeping your body healthy is a great way to keep your attitude healthy!

I'd like to close with a quote from Pulitzer Prize–winning novelist Ellen Glasgow that sums up the themes of this chapter.

In the past few years, I have made a thrilling discovery . . . that until one is over sixty, one can never really learn the secret of living. One can then begin to live, not simply with the intense part of oneself, but with one's entire being.

8

Creativity and Aging

Creativity represents a miraculous coming together of the unin-
hibited energy of the child with its apparent opposite and enemy:
the sense of order imposed by the disciplined adult intelligence.

—Norman Podhoretz

ON FEBRUARY 21, 1983, the largest television audience recorded to
date watched the final episode of $M^*A^*S^*H$, the long-running
series about an army surgical unit in the Korean War. Alan Alda,
the star of the show, also wrote or directed many of the episodes.
Over the course of the series, he won Emmy Awards for acting,
writing, and directing—the only person ever to do so.

Alda was forty-seven when the finale aired, and he could easily have retired. Instead, he pushed in new directions, working for
a time with Woody Allen and taking roles that worked against the
"type" he had established on $M^*A^*S^*H$. Some of the films
bombed at the box office, and critics were sometimes harsh in their

reviews of his acting or directing. Nonetheless, he pushed on, sometimes taking on jobs purely because he was personally fascinated by the subject, such as hosting the PBS science series *Scientific American Frontiers.*

By the time this book is in print, Alda will be seventy—and he continues to push his creative envelope. In 2004 he was nominated for an Oscar for his role as Senator Brewster in *The Aviator,* and in 2005 he was nominated for a Tony Award for his acting in David Mamet's Broadway play *Glengarry Glen Ross.* Talking about working on a Mamet play, Alda reveals just how close he still is to the cutting edge of his own creativity.

"The unconscious is darting left and right in Mamet's talk," he said in a recent interview in the *New Yorker.* "Nobody can come in and just do their part. Just as a chamber-music group will give an entirely different show from night to night, the play comes out with different colors and flavors each time. It's exciting to stay within its technical limits and still find a whole universe of difference."

While most of us accept that wisdom is a special provenance of aging, many people have the reverse view of creativity: they believe it is a flower of youth that blooms less and less frequently as the decades pass. This is yet another myth about aging that persists stubbornly in the face of overwhelming evidence to the contrary. Indeed, creativity is a happy potential in all ages, and, as Norman Podhoretz's words at the start of this chapter suggest, it can deepen and become richer with age.

My previous book, *The Creative Age: Awakening Human Potential in the Second Half of Life,* documented the depth and breadth of creative potential and expression throughout the second half of life. And in this book we've seen many examples of creativity in older adults, from the ingenious pizza delivery solution of my in-laws stuck in a snowstorm to the marvelous ideas for a September 11 memorial by Donal McLaughlin to the "First Hundred Years" recipe book put together by centenarian Anna Franklin.

Creative expression in the second half of life is fueled by the drives and desires of the Inner Push. When I use the word "creativity" I don't mean simply talents such as writing, painting, sculpture, or musical composition. I think we can all be creative in our own ways, whether we are artists or assembly-line workers, pianists or plumbers. Creativity can emerge in any realm, from the most abstract fields of science to the most intimate circles of human relations. The point is that creativity is the process of bringing something new into existence—and novelty is everywhere you look.

You can define creativity many ways, of course. Psychologist Howard Gardner, of Harvard University, differentiates between "big C" creativity and "little c" creativity. "Big C" applies to the extraordinary accomplishments of great artists, scientists, and inventors. These forms of creativity typically change fields of thought and the course of progress, as with Einstein's theory of relativity, Picasso's invention of cubism, and Edison's electrical inventions.

Creativity with a "little c" is grounded in the diversity of everyday activities and accomplishments. "Every person has certain areas in which he or she has a special interest," Gardner explains. "It could be something they do at work—the way they write memos or their craftsmanship at a factory—or the way they teach a lesson or sell something. After working at it for a while they can get to be pretty good—as good as anybody whom they know in their immediate world."

A common forum for "little c" creativity is gardening. Denise Driscoll, sixty-eight, a participant in my retirement study, prided herself on discovering exotic seeds in obscure plant catalogs. She would then use these "raw materials" in the gardens around her house, which she constantly modified for visual effect.

Sometimes a "little c" creative endeavor can leap to the "Big C" arena. For example, Maria Anne Smith, living in nineteenth-century Australia, was a respected grower of fruits. Like Denise Driscoll, Maria Smith enjoyed looking for the unusual and using it in novel ways. One day when she was sixty-nine, she spotted a seedling rising from a pile of discarded French crabapples. It seemed different, so she transplanted and nurtured it. The plant turned out to be a natural mutant—called a "sport"—and, in this case, one with some appealing features. Over the years, Smith used cuttings from the tree to increase her harvest and sell saplings to others. The fruit of that tree is now world famous: the Granny Smith Apple.

Another way to look at creativity came to me in a dream. I have always been fascinated by the creative genius of Albert Einstein and his elegant equation describing the equivalence of energy (e) and matter (m): $e = mc^2$ (c stands for the speed of light). In my dream the equation rearranged itself into a creativity equation: $c = me^2$. In this case, c stands for creativity, m stands for a person's mass of knowledge, and e stands for experience. The equation says that our creativity equals our mass of knowledge multiplied by the effects of our experience, which must be considered in two specific dimensions, inner (psychological and emotional) and outer (accumulated life experience, understanding, and perspective). This playful equation suggests that creativity is a function of both knowledge and experience, both of which increase with age.

DIFFERENT CATEGORIES OF CREATIVITY WITH AGING

I've found that creativity in the second half of life follows three basic patterns:

- Commencing creativity
- Continuing or changing creativity
- Creativity connected with loss

Commencing Creativity

Some people first significantly tap into their creative potential around age sixty-five. My own ideas about such "late bloomers" blossomed after a visit to a retrospective exhibit of a half century of folk art at Washington's Corcoran Gallery of Art. The works of twenty of the best African American folk artists from 1930 to 1980 were exhibited. Reading the artists' brief biographies, I discovered that of the twenty exhibitors, sixteen—80 percent—had begun painting or reached a recognizable mature phase as artists after the age of sixty-five. *Thirty percent* of them were eighty years of age or older.

After seeing the show I carried out a more formal study of folk art in the United States and found that across the ethnic and racial diversity of our society, folk art has been dominated by older adults.

Many of these individuals were finally free to pursue their artistic interests only after relinquishing other responsibilities. I saw their abilities as the outgrowth of their new external freedom from previous commitments combined with the inner freedom of the liberation phase attributes of experimentation and a desire to try new things. That an entire field can be dominated by older people makes an important and concrete statement about the depth of creative potential with aging.

My Aunt Esther Grushka exemplified the ageless quality of creativity. After devoting her life to her family and husband, with whom she ran a chain of retail stores, she felt a desire to do her own thing well up inside her. In her school days, Esther's teachers told her she had artistic talent, but she never followed up on it. Then, in her sixties, with her husband's businesses scaled back and her children grown, Esther began to draw and paint. One day on a visit to my parents I was amazed to see a painting of me on the living room wall. Aunt Esther had used my high school graduation picture as the model. I'd always hated that photograph, but Esther's brush had transformed it into something I took pleasure in. For the next twenty years, Esther continued to paint, and although her works will never hang in a gallery, they served as a fountainhead of satisfaction and pleasure in her later years.

Continuing or Changing Creativity

Some people find a creative outlet early in their lives and stick with it, often building careers around their talent. For such people, entering the second half of life and passing through the phases of aging can catalyze new creative expression.

This was certainly the case for Herbert Block, aka "Herblock," whose nationally syndicated political cartoons informed and enriched our culture for more than seventy years. His first cartoon appeared when he was in his twenties; his last was published less than two months before he died at ninety-one. Block was in his early sixties when he began to pointedly depict the unfolding Watergate fiasco. At this point he had the self-confidence to buck the editorial opinions of his home paper, the *Washington Post*, which, despite its own reporting, supported President Nixon in the early months of the investigation. Eventually, of course, the *Post* came around to the view that Block had been advocating early on. I see this as an example of creativity that is strengthened and extended by the developmental intelligence of aging.

Block concluded his autobiography with a revealing statement about the enduring potency of creativity: "There's always a clean slate, a fresh sheet of paper, a waiting space, a chance to have another shot at it tomorrow."

People also often experience a change in their creativity as they move through the four phases of later life. The midlife reevaluation phase may push them in new directions, such as the experimental roles that Alan Alda has pursued in the later phase of his career. The liberation phase can grant us a new sense of inner freedom, whereas the summing up phase can motivate us to deal creatively with unfinished business, or it can create a desire to express autobiographical impulses. The widening and deepening perspective in the encore phase can result in a similar shift of creative focus. The great mathematician and philosopher Bertrand Russell, for instance, focused tightly on mathematics in his youth and middle age. When he was forty-two, he and Alfred North Whitehead published the *Principia Mathematica*, which remains a masterpiece of mathematical logic and synthesis. As he grew older, his focus shifted to deeper issues, particularly philosophy and the many social ills of our time. At the age of seventy-three, he published his renowned work, *A History of Western Philosophy*, and he remained passionately involved in issues of peace and justice until he died at ninety-eight. In his autobiography, published just a year before his death, he wrote:

> I have lived in the pursuit of a vision, both personal and social. Personal: to care for what is noble, for what is beautiful, for what is gentle; to allow moments of insight to give wisdom at more mundane times. Social: to see in imagination the society that is to be created, where individuals grow freely, and where hate and greed and envy die because there is nothing to nourish them.

At the "little c" level in changing creativity, an acquaintance of mine, Art Reynolds, provides an example. Art spent his career designing computer software, for which he was both well paid and much praised. But when he turned sixty, Art told me he was feeling the Inner Push for change. "My work has always been for others—what *they* wanted," he said.

He decided to partially retire and pursue his interest in photography and the then-new field of computer design. This allowed him to combine a standing love of the visual arts with his technical expertise. The images he produced were striking, and about a year and a half after he began, he was invited to display them at an exhibit of avant-garde photography at a local gallery.

Creativity Connected with Loss

Aging is associated with a range of losses to which we must adapt. Creative outlets can help us cope and even transcend such loss. In fact, experiencing loss often triggers a creative response. We may, in the process, discover talents or skills we didn't know we had or which we had underestimated.

The life and work of William Carlos Williams illustrates this relationship. Williams was both a great poet and a respected pediatrician. When he was in his sixties, he suffered a stroke that impaired his motor abilities but left his intellect intact. He gave up medicine, a loss that threw him into a deep depression. It took hospitalization and several years for him to recover from the trauma, but after his recovery he experienced a torrent of creativity in his seventies. When he was seventy-nine, he published *Pictures from Bruegel,* which was awarded a Pulitzer Prize. In his later life, Williams wrote of an "old age that adds as it takes away," which is exactly what his later life epitomizes.

I experienced this phenomenon myself. As I noted in chapter 3, when I was in my late forties I was misdiagnosed as having Lou Gehrig's disease. In a desperate attempt to deal with the overwhelming emotions I was feeling, I started developing games for older adults. This crisis and sense of impending profound loss (aided also by being in the midlife reevaluation phase) stimulated skills I didn't know I had and launched me into a whole new phase of creative potential.

Creativity's Positive Impact on Health

Creativity in later life is not just an academic subject or an "ornament" of life—nice, but not really necessary. A great deal of research, both psychological and physiological, has demonstrated that creativity is good for one's health. And it's not just that people who are engaged creatively are healthier to begin with. The very act of engaging one's mind in creative ways affects health directly via the many mind/body connections. Our brains are deeply connected to our bodies via nerves, hormones, and the immune system. Anything that stimulates the brain, reduces stress, and promotes a more balanced emotional response will trigger positive changes in the body. Some preliminary findings show that sustained creativity enhances recovery from infections and injuries as well as reducing the pain or discomfort of chronic conditions such as arthritis. The explosion of so many types of art-related therapy—dance and movement therapy, music therapy, poetry therapy, drama therapy, and visual arts therapy—reflect this research.

My own contribution to this field of scientific research began in 2001 with a grant from the National Endowment for the Arts, in coordination with five other sponsors, to conduct a rigorous national study examining the effects of community-based art programs on the health and functioning of older adults. The study compares the physical and mental health and the social functioning of 150 older persons involved in the arts programs to a comparable group of 150 adults who are not in such programs (the control group). All the participants are age sixty-five or older, most were living independently when the study began, and the two groups were comparable in their health and functioning at the start of the study. The adults who were not in the arts group were free to socialize, attend classes, or do any of their normal activities, including art (although none in the control group became involved in rigorous and sustained participatory art programs). We wanted to try, as best we could, to see if it was the creativity involved in the arts programs that made a difference rather than the mere fact that the participants were engaged in a regular, structured social situation.

The arts groups met for thirty-five weekly meetings—analogous to a college course. There were also between-session assignments as well as exhibitions and concerts. For example, a chorale at one site gave some ten concerts a year in addition to their regular weekly practice sessions.

We assessed each person's health and social functioning with comprehensive questionnaires at the beginning of the programs, at the halfway mark, and at the end, two years after starting. Our hypothesis was that the people who participated in the arts programs would show less decline than the control group, who did not participate in those programs. We were pleasantly surprised, therefore, when the initial results exceeded our expectations. Many people in the arts groups had *stabilized* their health—meaning they didn't decline at all—and some actually *improved* their health. This in a group of people with an average age of eighty, which is greater than the current life expectancy!

Here are the major findings from the first phase of the study, which was conducted in the Washington, D.C., area under the artistic direction of Jeanne Kelly with the Levine School of Music. (Similar paired study groups are being researched in Brooklyn at

Elders Share the Arts, under the direction of Susan Perlstein, and in San Francisco at the Center for Elders and Youth in the Arts, under the direction of Jeff Chapline.) All the results were statistically significant, meaning that they reflected real differences between the two study groups. Compared with the control group, those who participated in the community arts program:

- Had better health after one year (those in the control group reported that their health was not as good after the same elapsed time)
- Had fewer doctor visits (although both groups had more visits compared with a year earlier)
- Used fewer medications
- Felt less depressed
- Were less lonely
- Had higher morale
- Were more socially active

These remarkable preliminary results have attracted attention in both scientific and lay circles. Clearly the community-based art programs are having a real effect on health promotion and disease prevention, which in turn support the independence of older individuals and their ability to live in their communities.

EXPLAINING THE RESULTS

The positive results of this study are explained by three key factors or mechanisms.

1. *Sense of control.* An important body of research on aging known as "sense of control studies" shows that older persons engaged in activities in which they experience a sense of mastery and control have better health than those who don't. Interestingly, this influence becomes stronger with age. The arts programs at the heart of our study gave participants a mounting sense of control and mastery. Each new week was like a booster shot, as evidenced by the following quotes from participants:

"I had no idea I could read music this well, and actually improve!"

"My wife loves the jewelry I make for her—it's been great for our marriage."

"I can't believe I write as well as I do."

"Several of my friends came to the art exhibition that included three pieces of my work; it really made me feel good."

"My grandson said, 'Grandma, you're a poet who doesn't know it.'"

Gaining a sense of mastery in one area can lead to feelings of empowerment that spread to other spheres of life, leading to more confidence, a willingness to take risks, and the energy for trying new things. This helps explain why those in the art programs were more socially engaged after a year than those in the control group, who had a *reduction* in their overall activities.

Enhancing one's sense of control and mastery can also directly promote physical health. Researchers in the field of psychoneuroimmunology, which links psychology with immunology, have found that the positive feelings associated with a sense of control boost the immune system. Specifically, a sense of well-being appears to stimulate the production of important immune system cells such as white blood cells and the so-called natural killer cells that attack tumor cells and infected body cells.

2. *Social engagement.* The second major factor behind the positive results of our study is the social engagement fostered by the structured art activities. Again, evidence from psychoneuroimmunology suggests that relating to people, forming or maintaining strong friendships, and being involved in a variety of social settings are all associated with better health and reduced mortality. For example, having active social relationships in the second half of life has been associated with reduced blood pressure. Having good relationships is also associated with lower stress levels, which helps preserve the integrity of the immune system.

One of the striking observations made of those participating in the art groups was the degree to which group members supported each other in the face of loss or difficulty. For example, when one of the women in our Brooklyn writing group was in the hospital, several of the members visited her, brought her assignments, discussed her work, and brought her work back to the group for further feedback.

"My spirits were so lifted," she said. "I'm sure it speeded my recovery. I couldn't wait to get back into the swing of things and share another essay with the whole group."

3. *The engaging nature of art.* To be effective over the long haul, any health-promoting activity must be sustained. This can be difficult when the activity is inherently boring (such as walking on a treadmill) or unpleasant (hunger pangs from dieting or aches from exercise, for example). Fortunately, the nature of creative

endeavors makes them self-sustaining. They are almost always interesting, stimulating, and pleasurable. The fact that they also provide a venue for individual mastery and social engagement makes community-based art programs a "package deal" for health promotion. And they are easier to implement and more readily available than other options: most localities have the resources to start a community-based art program. (For information about setting up such a program, see the entry for the Elders Share the Arts program in appendix 2.)

The engaging nature of art promotes sustained involvement, and, as we saw in chapter 6, activities that repeat over time are more valuable than one-shot experiences. The creativity study bore this out. Many people in the control group were just as "busy" as those participating in the art programs. But simply being involved in many transient activities with limited potential for fostering mastery or building new relationships did not translate into improvement in health. It wasn't just the level of activity but the *nature* of that activity that made a difference.

A conversation I had with one of the participants in the Washington, D.C., group illustrates the point. Margaret Spencer was ninety-four when she got involved in a chorale as part of the program. She was hesitant to join because she didn't know if she could read music quickly enough. But, prodded by others, she joined.

"I was amazed that I could sing, better than I ever thought, *and* I could keep up," she told me. "I'm even getting better! I'm ninety-four and I'm making new friends. I want to continue with this."

To summarize: Any kind of health-promotion program should create opportunities to achieve individual mastery and sense of control; support social engagement and relationship building; and have built-in sustainability over the long haul.

Most fundamentally, this study, like this book as a whole, illustrates the good that can come from controlling our own choices and our own destiny. We can, if we want to, learn, grow, love, and experience profound happiness in our later years. We need not succumb to difficulties, nor need we accept the myths that still exist about growing older. Nature endows us with the mental potential we need to improve our health, our relationships, and our lives. I hope this book and the stories I have told in it encourage you to begin—today—to make a positive difference in your own life.

To exist is to change, to change is to mature,
to mature is to go on creating oneself endlessly.
—Henri Bergson

Appendix 1: Notes

Introduction

Moving from a Problem Focus to a Potential Focus with Aging
Cohen, G. D. 2000. *The Creative Age: Awakening Human Potential in the Second Half of Life*. New York: Avon.

The Concept of Successful Aging
Rowe, J. W., and R. L. Kahn. 1998. *Successful Aging*. New York: Pantheon Books.

Trends in the Health of Older Americans
Pastor, P. N., D. M. Makuc, C. Reuben, and H. Xia. 2002. *Chartbook on Trends in the Health of Americans*. Hyattsville, MD: National Center for Health Statistics. Available at http://www.cdc.gov/nchs/data/hus/hus02.pdf.

CHAPTER 1
THE POWER OF OLDER MINDS

Adverse Effects of Stress on Brain Structures and Neurogenesis

Gould, E., and P. Tanapat. Stress and hippocampal neurogenesis. *Biological Psychiatry* (1999) 46(11): 1472–1479.

Bilateral Involvement of the Brain's Two Hemispheres:
The HAROLD Model

Cabeza, R. Hemispheric asymmetry reduction in older adults: The HAROLD model. *Psychology and Aging* (2002) 17(1): 85–100.

Cabeza, R., N. D. Anderson, J. K. Locantore, and A. R. McIntosh. Aging gracefully: Compensatory brain activity in high-performing older adults. *Neuroimage* (2002) 17(3): 1394–1402.

Hazlett, E. A., M. S. Buchsbaum, R. C. Mohs, J. Spiegel-Cohen, T. C. Wei, R. Azueta, M. M. Haznedar, M. B. Singer, L. Shihabuddin, C. Luu-Hsia, and P. D. Harvey. Age-related shift in brain region activity during successful memory performance. *Neurobiology of Aging* (1998) 19(5): 437–445.

Pujol, J., A. Lopez-Sala, J. Deus, N. Cardoner, N. Sebastian-Galles, G. Conesa, and A. Capdevila. The lateral asymmetry of the human brain studied by volumetric magnetic resonance imaging. *Neuroimage* (2002) 17(2): 670–679.

Reuter-Lorenz, P. A., J. Jonides, E. E. Smith, A. Hartley, A. Miller, C. Marshuetz, and R. A. Koeppe. Age differences in the frontal lateralization of verbal and spatial working memory revealed by PET. *Cognitive Neuroscience* (2000) 12(1): 174–187.

Appendix 1: Notes

Brain Recruitment with Aging

Gunning-Dixon, F. M., R. C. Gur, A. C. Perkins, L. Schroeder, T. Turner, B. I. Turetsky, R. M. Chan, J. W. Loughead, D. C. Alsop, J. Maldjian, and R. E. Gur. Age-related differences in brain activation during emotional face processing. *Neurobiology of Aging* (2003) 24(2): 285–295.

Langenecker, S. A., and K. A. Nielson. Frontal recruitment during response inhibition in older adults replicated with MRI. *Neuroimage* (2003) 20(2): 1384–1392.

Brain Remodeling and Aging

Biegler, R., A. McGregor, J. R. Krebs, and S. D. Healy. A larger hippocampus is associated with longer-lasting spatial memory. *Proceedings of the National Academy of Sciences of the United States of America* (2001) 98(12): 6941–6944.

Geinisman, Y., J. F. Disterhoft, H. J. G. Gundersen, M. D. McEchron, I. S. Persina, J. M. Power, E. A. Van der Zee, and M. D. West. Remodeling of hippocampal synapses following hippocampus-dependent associative learning. *Journal of Comparative Neurology* (in press).

Maguire, E. A., R. S. Frackowiak, and C. D. Frith. Recalling routes around London: Activation of the right hippocampus in taxi drivers. *Journal of Neuroscience* (1997) 17(18): 7103–7110.

Maguire, E. A., D. G. Gadian, I. S. Johnsrude, C. D. Good, J. Ashburner, R. S. J. Frackowiak, and C. D. Frith. Navigation-related structural change in the hippocampi of taxi drivers. *Proceedings of the National Academy of Sciences of the United States of America* (2000) 97(8): 4398–4403.

Maguire, E. A., H. J. Spiers, C. D. Good, T. Hartley, R. S. Frackowiak, and N. Burgess. Navigation expertise and the human hippocampus: A structural brain imaging analysis. *Hippocampus* (2003) 13(2): 250–259.

Cardiovascular Fitness and Cortical Plasticity with Aging
Colcombe, S. J., A. F. Kramer, K. I. Erickson, P. Scalf, E. McAuley, N. J. Cohen, A. Webb, G. J. Jerome, D. X. Marquez, and S. Elavsky. Cardiovascular fitness, cortical plasticity, and aging. *Proceedings of the National Academy of Sciences of the United States of America* (2004) 101(9): 3316–3321.

Colcombe, S. J., A. F. Kramer, E. McAuley, K. I. Erickson, and P. Scalf. Neurocognitive aging and cardiovascular fitness: Recent findings and future directions. *Journal of Molecular Neuroscience* (2004) 24(1): 9–14.

Environmental Influences of Brain Plasticity with Aging
Kolb, B., and I. Q. Whishaw. Brain plasticity and behavior. *Annual Review of Psychology* (1998) 49: 43–64.

Kramer, A. F., L. Bherer, S. J. Colcombe, W. Dong, and W. T. Greenough. Environmental influences on cognitive and brain plasticity during aging. *Journal of Gerontology: Medical Sciences* (2004) 59A(9): 940–957.

Rossini, P. M., and G. Dal Forno. Integrated technology for evaluation of brain function and neural plasticity. *Physical Medicine and Rehabilitation Clinics of North America* (2004) 15(1): 263–306.

Games and Leisure Activities Protecting Against Dementia
Coyle, J. T. Use it or lose it: Do effortful mental activities protect against dementia? *New England Journal of Medicine* (2003) 348(25): 2489–2490.

Verghese, J., R. B. Lipton, M. J. Katz, C. B. Hall, C. A. Derby, G. Kuslansky, A. F. Ambrose, M. Sliwinski, and H. Buschke. Leisure activities and the risk of dementia in the elderly. *New England Journal of Medicine* (2003) 348(25): 2508–2516.

Neurogenesis and Aging

Alvarez-Buylla, A., and J. M. García-Verdugo. Neurogenesis in adult subventricular zone. *Journal of Neuroscience* (2002) 22(3): 629–634.

Derrick, B. E., A. D. York, and J. L. Martinez Jr. Increased granule cell neurogenesis in the adult dentate gyrus following mossy fiber stimulation sufficient to induce long-term potentiation. *Brain Research* (2000) 857(1–2): 300–307.

Gould, E., and C. G. Gross. Neurogenesis in adult mammals: Some progress and problems. *Journal of Neuroscience* (2002) 22(3): 619–623.

Kempermann, G., L. Wiskott, and F. H. Gage. Functional significance of adult neurogenesis. *Current Opinion in Neurobiology* (2004) 14(2): 186–191.

Kintner, C. Neurogenesis in embryos and in adult neural stem cells. *Journal of Neuroscience* (2002) 22(3): 639–643.

Nottebohm, F. Why are some neurons replaced in adult brain? *Journal of Neuroscience* (2002) 22(3): 624–628.

Rakic, P. Adult neurogenesis in mammals: An identity crisis. *Journal of Neuroscience* (2002) 22(3): 614–618.

Santarelli, L., M. Saxe, C. Gross, A. Surget, F. Battaglia, S. Dulawa, N. Weisstaub, J. Lee, R. Duman, O. Arancio, C. Belzung, and R. Hen. Requirement of hippocampal neurogenesis for the behavioral effects of antidepressants. *Science* (2003) 301(5634): 805–809.

Taupin, P., and F. H. Gage. Mini-review: Adult neurogenesis and neural stem cells of the central nervous system in mammals. *Journal of Neuroscience Research* (2002) 69: 745–749.

Physical Activity Enhancing Cognitive Functioning with Aging

Dik, M., D. J. Deeg, M. Visser, and C. Jonker. Early life physical activity and cognition at old age. *Journal of Clinical and Experimental Neuropsychology* (2003) 25(5): 643–653.

Laurin, D., R. Verreault, J. Lindsay, K. MacPherson, and K. Rockwood. Physical activity and risk of cognitive impairment and dementia in elderly persons. *Archives of Neurology* (2001) 58(3): 498–504.

Yaffe, K., D. Barnes, M. Nevitt, L. Y. Lui, and K. Covinsky. A prospective study of physical activity and cognitive decline in elderly women: Women who walk. *Archives of Internal Medicine* (2001) 161(14): 1703–1708.

Rehabilitation Inducing Brain Plasticity in Stroke Patients

Fraser, C., M. Power, S. Hamdy, J. Rothwell, D. Hobday, I. Hollander, P. Tyrell, A. Hobson, S. Williams, and D. Thompson. Driving plasticity in human adult motor cortex is associated with improved motor function after brain injury. *Neuron* (2002) 34(5): 831–840.

Horner, P. J., and F. Gage. Regeneration in the adult and aging brain. *Archives of Neurology* (2002) 59: 1717–1720.

Johansson, B. B. Brain plasticity in health and disease. *Keio Journal of Medicine* (2004) 53(4): 231–246.

Nudo, R. J. Adapative plasticity in motor cortex: Implications for rehabilitation after brain injury. *Journal of Rehabilitation Medicine* (2003) 41S:7–10.

Papathanasiou, I., S. R. Filipovic, R. Whurr, and M. Jahanshahi. Plasticity of motor cortex excitability induced by rehabilitation therapy for writing. *Neurology* (2003) 61(7): 881–882.

Synapses and Brain Plasticity

LeDoux, J. 2002. *Synaptic Self: How Our Brains Become Who We Are.* New York: Viking.

Nakamura, H., S. Kobayashi, Y. Ohashi, and S. Ando. Age-changes of brain synapses and synaptic plasticity in response to an enriched environment. *Journal of Neuroscience Research* (1999) 56(3): 307–315.

Trachtenberg, J. T., B. E. Chen, G. W. Knott, G. Feng, J. R. Sanes, E. Welker, and K. Svoboda. Long-term in vivo imaging of experience-dependent synaptic plasticity in adult cortex. *Nature* (2002) 420(6917): 788–794.

Chapter 2
Harnessing Developmental Intelligence

College Graduates Over Thirty and Forty Years of Age
Bradburn, E. M., R. Berger, X. Li, K. Peter, and K. Rooney. A descriptive summary of 1999–2000 bachelor's degree recipients one year later: With an analysis of time to degree. *Education Statistics Quarterly* (2003) 5(3): 96–158.

Psychological Development Across the Life Cycle
Baltes, P. B., U. M. Staudinger, and U. Lindenberger. Lifespan psychology: Theory and application to intellectual functioning. *Annual Review of Psychology* (1999) 50: 471–507.

Colarusso, C. A., and R. A. Nemiroff. 1981. *Adult Development*. New York: Plenum Press.

Erikson, E. E. 1980. *Identity and the Life Cycle*. New York: W. W. Norton.

Jones, C. J., and W. Meredith. Developmental paths of psychological health from early adolescence to later adulthood. *Psychology and Aging* (2000) 15(2): 351–360.

Pinker, S. 2002. *The Blank Slate: The Modern Denial of Human Nature*. New York: Penguin.

Sheldon, K. M., and T. Kasser. Getting older, getting better? Personal strivings and psychological maturity across the life span. *Developmental Psychology* (2001) 4: 491–501.

Vaillant, George. 2002. *Aging Well*. New York: Little, Brown and Company.

Psychotherapy Influencing Brain Plasticity
Kandel, E. R. Biology and the future of psychoanalysis: A new intellectual framework for psychiatry revisited. *American Journal of Psychiatry* (1999) 156: 505–524.

CHAPTER 3
THE SECOND HALF OF LIFE: PHASES I AND II

A New Theory of Psychological Development in the Second Half of Life
Cohen, G. D. Human potential phases in the second half of life: Mental health theory development. *American Journal of Geriatric Psychiatry* (1999) 7(1): 1–7.

Cohen, G. D. 2000. *The Creative Age: Awakening Human Potential in the Second Half of Life*. New York: Avon.

Cohen, G. D. Creativity with aging: Four phases of potential in the second half of life. *Geriatrics* (2001) 56(4): 51–57.

Cohen, G. D. 2004. *Using the Heart and the Mind: Human Development in the Second Half of Life*. Mind Alert. San Francisco: American Society on Aging.

Dendritic Density and Length During the Liberation Phase
Flood, D. G., S. J. Buell, C. H. Defiore, G. J. Horwitz, and P. D. Coleman. Age related dendritic growth in dentate gyrus of human brain is followed by regression in the "oldest old." *Brain Research* (1985) 345(2): 366–368.

Flood, D. G., and P. D. Coleman. Hippocampal plasticity in normal aging and decreased plasticity in Alzheimer's disease. *Progress in Brain Research* (1990) 83: 435–443.

Neurogenesis in the Aging Hippocampus: Relevance to Novelty
Kempermann, G. Why new neurons? Possible functions for adult hippocampal neurogenesis. *Journal of Neuroscience* (2002) 22(3): 635–638.

Kempermann, G., and L. Wiskott. 2003. What is the functional role of new neurons in the adult dentate gyrus? In *Stem Cells in the Nervous System: Function and Clinical Implications*, edited by F. H. Gage, A. Bjorklund, A. Prochiantz, and Y. Christen. Berlin: Springer, 57–65.

Ribak, C. E., M. J. Korn, Z. Shan, and A. Obenaus. Dendritic growth cones and recurrent basal dendrites are typical features of newly generated dentate granule cells in the adult hippocampus. *Brain Research* (2004) 1000(1–2): 195–199.

Van Praag, H., A. F. Schinder, B. R. Christie, N. Toni, T. D. Palmer, and F. II. Gage. Functional neurogenesis in the adult hippocampus. *Nature* (2002) 415(6875): 1030–1034.

Positive Changes Connected with Developmental Intelligence
Stewart, A. J., J. M. Ostrove, and R. Helson. Middle aging in women: Patterns of personality change from the 30s to the 50s. *Journal of Adult Development* (2001) 8(1): 23–37.

CHAPTER 4
THE SECOND HALF OF LIFE: PHASES III AND IV

Bilateral Hippocampal Involvement and Autobiography with Aging
Maguire, E. A., and C. D. Frith. Aging affects the engagement of the hippocampus during autobiographical memory retrieval. *Brain: A Journal of Neurology* (2003) 126(7): 1511–1523.

Ryan, L., L. Nadel, K. Keil, K. Putnam, D. Schnyer, T. Trouard, and M. Moscovitch. Hippocampal complex and retrieval of recent and very remote autobiographical memories: Evidence from functional magnetic resonance imaging in neurologically intact people. *Hippocampus* (2001) 11(6): 707–714.

Brain Volume in the Eleventh Decade
Mueller, E. A., M. M. Moore, D. C. Kerr, G. Sexton, R. M. Camicioli, D. B. Howieson, J. F. Quinn, and J. A. Kaye. Brain volume preserved in healthy elderly through the eleventh decade. *Neurology* (1998) 51(6): 1555–1562.

Centenarians
Ellis, N. 2002. *If I live to Be 100*. New York: Three Rivers.

Perls, T. T. 1999. *Living to 100: Lessons in Living to Your Maximum Potential at Any Age.* New York: Basic Books.

Life Satisfaction and Positive Emotions at an Advanced Age

Hamarat, E., D. Thompson, F. Aysan, D. Steele, K. Matheny, and C. Simons. Age differences in coping resources and satisfaction with life among middle-aged, young-old, and oldest-old adults. *Journal of Genetic Psychology* (2002) 163(3): 360–367.

Isaacowitz, D. M., G. E. Vaillant, and M. E. Seligman. Strengths and satisfaction across the adult lifespan. *International Journal of Aging and Human Development* (2003) 57(2): 181–201.

Leigland, L. A., L. E. Schulz, and J. S. Janowsky. Age related changes in emotional memory. *Neurobiology of Aging* (2004) 25(8): 1117–1124.

Mather, M., T. Canli, T. English, S. Whitfield, P. Wais, K. Ochsner, J. D. Gabrieli, and L. L. Carstensen. Amygdala responses to emotionally valenced stimuli in older and younger adults. *Psychological Science* (2004) 15(4): 259–263.

Pasupathi, M., and L. L. Carstensen. Age and emotional experience during mutual reminiscing. *Psychology and Aging* (2003) 18(3): 430–442.

CHAPTER 5
COGNITION, MEMORY, AND WISDOM

Brain Repair with Aging

Brazel, C. Y., and M. S. Rao. Aging and neuronal replacement. *Ageing Research Reviews* (2000) 3(4): 465–483.

Appendix 1: Notes

Brain Reserve with Aging

Cabeza, R., S. M. Daselaar, F. Dolcos, S. E. Prince, M. Budde, and L. Nyberg. Task-independent and task-specific age effects on brain activity during working memory, visual attention, and episodic retrieval. *Cerebral Cortex* (2004) 14(4): 364–375.

Chapman, P. F. Cognitive aging: Recapturing the excitation of youth? *Current Biology* (2005) (15)1: R31–R33.

Reutter-Lorenz, P. A., L. Stanczak, and A. C. Miller. Neural recruitment and cognitive aging: Two hemispheres are better than one, especially as you age. *Psychological Science* (1999) 10(6): 494–500.

Trollor, J. N., and M. J. Valenzuela. Brain ageing in the new millennium. *Australian and New Zealand Journal of Psychiatry* (2001) 35(6): 788–805.

Whalley, L. J., I. J. Deary, C. L. Appleton, and J. M. Starr. Cognitive reserve in the neurobiology of cognitive aging. *Ageing Research Reviews* (2004) 3(4): 369–382.

Dendritic Spines and Memory Storage

Farris, S. M., G. E. Robinson, and S. E. Fahrbach. Experience- and age-related outgrowth of intrinsic neurons in the mushroom bodies of the adult worker honeybee. *Journal of Neuroscience* (2001) 21(16): 6395–6404.

Grutzendler, J., N. Kasthuri, and W. B. Gan. Long-term dendritic spine stability in the adult cortex. *Nature* (2002) 420(6917): 812–816.

Halpain, S., K. Spencer, and S. Graber. Dynamics and pathology of dendritic spines. *Progress in Brain Research* (2005) 147: 29–37.

Leuner, B., J. Falduto, and T. J. Shors. Associative memory formation increases the observation of dendritic spines in the hippocampus. *Journal of Neuroscience* (2003) 23(2): 659–665.

Appendix 1: Notes

Developmental Intelligence
Labouvie-Vief, G., and M. Diehl. Cognitive complexity and cognitive-affective integration: Related or separate domains of adult development? *Psychology and Aging* (2000) 15(3): 490–504.

Luna, B., K. R. Thulborn, D. P. Munoz, E. P. Merriam, K. E. Garver, N. J. Minshew, M. S. Keshavan, C. R. Genovese, W. F. Eddy, and J. A. Sweeney. Maturation of widely distributed brain function subserves cognitive development. *Neuroimage* (2001) 13(5): 786–793.

Moody, H. R. 2003. Conscious aging: A strategy for positive change in later life. In *Mental Wellness in Aging,* edited by J. L. Ronch and J. A. Goldfield. Baltimore: Health Professions Press, 139–160.

Glass of Wine a Day Good for Cognition with Aging
Stampfer, M. J., J. H. Kang, J. Chen, R. Cherry, and F. Grodstein. Effects of moderate alcohol consumption on cognitive function in women. *New England Journal of Medicine* (2005) 352(3): 245–253.

Maintaining Memories with Aging
Bailey, C. H., E. R. Kandel, and K. Si. The persistence of long-term memory: A molecular approach to self-sustaining changes in learning-induced synaptic growth. *Neuron* (2004) 44(1): 49–57.

Cabeza, R., J. K. Locantore, and N. D. Anderson. Lateralization of prefrontal activity during episodic memory retrieval: Evidence for the production-monitoring hypothesis. *Journal of Cognitive Neuroscience* (2003) 15(2): 249–259.

Charles, S. T., M. Mather, and L. L. Carstensen. Aging and emotional memory: The forgettable nature of negative images for older adults. *Journal of Experimental Psychology* (2003) 132(2): 310–324.

Denburg, N. L., T. W. Buchanan, D. Tranel, and R. Adolphs. Evidence for preserved emotional memory in normal older persons. *Emotion* (2003) 3(3): 239–253.

Iaria, G., M. Petrides, A. Dagher, B. Pike, and V. D. Bohbot. Cognitive strategies dependent on the hippocampus and caudate nucleus in human navigation: Variability and change with practice. *Journal of Neuroscience* (2003) 23(13): 5945–5952.

Martin, S. J., P. D. Grimwood, and R. G. Morris. Synaptic plasticity and memory: An evaluation of the hypothesis. *Annual Review of Neuroscience* (2000) 23: 649–711.

Morcom, A. M., C. D. Good, R. S. Frackowiak, and M. D. Rugg. Age effects on the neural correlates of successful memory encoding. *Brain: A Journal of Neurology* (2003) 126(1): 213–229.

Park, D. C., R. C. Welsh, C. Marshuetz, A. H. Gutchess, J. Mikels, T. A. Polk, D. C. Noll, and S. F. Taylor. Working memory for complex scenes: Age differences in frontal and hippocampal activations. *Journal of Cognitive Neuroscience* (2003) 15(8): 1122–1134.

Positive Effects of Meditation on Mental Processes with Aging
Lutz, A., L. L. Greischar, N. B. Rawlings, M. Ricard, and R. J. Davidson. Long-term meditators self-induce high-amplitude gamma synchrony during mental practice. *Proceedings of the National Academy of Sciences of the United States of America* (2004) 101(46): 16369–16373.

Myelination with Aging
Bartzokis, G., I. L. Cummings, D. Sultzer, V. W. Henderson, K. H. Nuechterlein, and J. Mintz. White matter structural integrity in healthy adults and patients with Alzheimer's disease: A magnetic resonance imaging study. *Archives of Neurology* (2003) 60(3): 393–398.

Sowell, E. R., P. M. Thompson, and A. W. Toga. Mapping changes in the human cortex throughout the span of life. *Neuroscientist* (2004) 10(4): 372–392.

Neurogenesis and Learning
Prickaerts, J., G. Koopmans, A. Blokland, and A. Scheepens. Learning and adult neurogenesis: Survival with or without proliferation? *Neurobiology of Learning and Memory* (2004) 81(1): 1–11.

Postformal Thought and Aging
Arlin, P. K. Cognitive development in adulthood: A fifth stage? *Developmental Psychology* (1975) 11: 602–606.

Marchand, H. Some reflections on postformal thought. *The Genetic Epistemologist* (2001) 29(3): 2–9.

Pliske, R. M., and S. A. Mutter. Age differences in the accuracy of confidence judgments. *Experimental Aging Research* (1996) 22(2): 199–216.

Richards, F., and M. Commons. 1990. Postformal cognitive-developmental theory and research: A review of its currents status. In *Higher Stages of Human Development,* edited by C. Alexander and E. Langer. New York: Oxford University Press, 139–161.

Sinnott, J. D. 1991. Limits to problem solving: Emotion, intention, goal, clarity, health, and other factors in postformal thought. In *Bridging Paradigms: Positive Development in Adulthood and Cognitive Aging,* edited by J. D. Sinnott and J. C. Cavanaugh. New York: Praeger.

Sinnott, J. D. 1999. Creativity and postformal thought: Why the last stage is the creative stage. In *Creativity and Successful Aging,* edited by C. E. Adams-Price. New York: Springer Publishing Company.

CHAPTER 6
CULTIVATING SOCIAL INTELLIGENCE

Broad Overview of Social Science Issues with Aging
Binstock, R. H., and L. K. George, editors. 2001. *Handbook of Aging and the Social Sciences,* 5th edition. New York: Academic Press.

Gender Differences in Social Roles and Issues with Aging
Arber, S., and H. Cooper. Gender differences in health in later life: The new paradox? *Social Science and Medicine* (1999) 48(1): 61–76.

Moen, P. 2001. The gendered life course. In *Handbook of Aging and the Social Sciences*, 5th edition, edited by R. H. Binstock and L. K. George. New York: Academic Press, 179–196.

Social Behavior Leading to a New Evolutionary View of the Relevance of Aging
Lee, R. D. Rethinking the evolutionary theory of aging: Transfers, not births, shape senescence in social species. *Proceedings of the National Academy of Sciences of the United States of America* (2003) 100(16): 9637–9642.

Social Environment Influencing Neurogenesis and Neuroplasticity with Aging
Lu, L., G. Bao, H. Chen, P. Xia, X. Fan, J. Zhang, G. Pei, and L. Ma. Modification of hippocampal neurogenesis and neuroplasticity by social environments. *Experimental Neurology* (2003) 183(2): 600–609.

Social Reasoning with Aging
Blanchard-Fields, F. Reasoning on social dilemmas varying in emotional saliency: An adult developmental perspective. *Psychology and Aging* (1986) 1(4): 325–333.

Blanchard-Field, F., and J. C. Irion. Coping strategies from the perspective of two developmental markers: Age and social reasoning. *Journal of Genetic Psychology* (1988) 149(2): 141–151.

Chapter 7
Reinventing Retirement

General Overview of Retirement
Cohen, G. D. Retirement: Advising older adults who are contemplating this change. *Geriatrics* (2002) 57(8): 37–38.

Henretta, J. C. 2001. Work and retirement. In *Handbook of Aging and the Social Sciences*, 5th edition, edited by R. H. Binstock and L. K. George. New York: Academic Press, 255–271.

Moen, P. 2000. *The Cornell Retirement and Well-Being Study: Final Report, 2000.* Ithaca, NY: Bronfenbrenner Life Course Center, Cornell University.

Learning Across the Life Cycle
Thornton, J. E. Life-span learning: A developmental perspective. *International Journal of Aging and Human Development* (2003) 57(1): 55–76.

Lifestyle and Cognitive Reserve with Aging
Fillit, H. M., R. N. Butler, A. W. O'Connell, M. S. Albert, J. E. Birren, C. W. Cotman, W. T. Greenough, P. E. Gold, A. F. Kramer, L. H. Kuller, T. T. Perls, B. G. Sahagan, and T. Tully. Achieving and maintaining cognitive vitality with aging. *Mayo Clinic Proceedings* (2002) 77(7): 681–696.

Scarmeas, N., and Y. Stern. Cognitive reserve and lifestyle. *Journal of Clinical and Experimental Neuropsychology* (2003) 25(5): 625–633.

Social Relations and Health with Aging

Avlund, K., M. T. Damsgaard, and B. E. Holstein. Social relations and mortality: An eleven year follow-up study of 70-year-old men and women in Denmark. *Social Science and Medicine* (1998) 47(5): 635–643.

Bennett, K. M. Low level social engagement as a precursor of mortality among people in later life. *Age and Ageing* (2002) 31(3): 165–168.

Glass, T. A., C. M. de Leon, R. A. Marottoli, and L. F. Berkman. Population based study of social and productive activities as predictors of survival among elderly Americans. *British Medical Journal* (1999) 319: 478–483.

The Experience Corps as a Social Model for Health Promotion

Fried, L. P., M. C. Carlson, M. Friedman, K. D. Frick, T. A. Glass, J. Hill, S. McGill, G. W. Rebok, T. Seeman, J. Tielsch, B. A. Wasik, and S. Zeger. A social model for health promotion for an aging population: Initial evidence on the Experience Corps model. *Journal of Urban Health: Bulletin of the New York Academy of Medicine* (2004): 81(1): 64–78.

The Social Portfolio

Cohen, G. D. Mental health promotion in later life: The case for the social portfolio. *American Journal of Geriatric Psychiatry* (1995) 3: 277–279.

Cohen, G. D. 2003. The social portfolio: The role of activity in mental wellness as people age. In *Mental Wellness in Aging*, edited by J. L. Ronch and J. A. Goldfield. Baltimore: Health Professions Press, 113–122.

CHAPTER 8
CREATIVITY AND AGING

Art Therapies for Older Adults
Cadigan, M. E., N. A. Caruso, S. M. Haldeman, M. E. McNamara, D. A. Noyes, M. A. Spadafora, and D. L. Carroll. The effects of music on cardiac patients on bed rest. *Progress in Cardiovascular Nursing* (2001) 16(1): 5–13.

Creativity and Aging Study
Cohen, G. D. National study documents benefits of creativity programs for older adults. *The Older Learner* (2005) 13(2): 1, 6.

Overview of Creativity and Aging
Adams-Price, C. E., editor. 1998. *Creativity and Successful Aging.* New York: Springer Publishing Company.

Cohen, G. D. Creativity and aging: Relevance to research, practice, and policy. *American Journal of Geriatric Psychiatry* (1994) 2(4): 277–281.

Cohen, G. D. Creativity with aging: Four phases of potential in the second half of life. *Geriatrics* (2001) 56(4): 51–57.

Practice Effects on Brain Plasticity Among Musicians
Ragert, P., A. Schmidt, E. Altenmuller, and H. R. Dinse. Superior tactile performance and learning in professional pianists: Evidence for meta-plasticity in musicians. *European Journal of Neuroscience* (2004) 19(2): 473–478.

Trainor, L. J., A. Shahin, and L. E. Roberts. Effects of musical training on the auditory cortex in children. *Annals of the New York Academy of Sciences* (2003) 999: 506–513.

Appendix 1: Notes

Psychoneuroimmunology and Health with Aging

Coe, C. L., and G. R. Lubach. Critical periods of special health relevance for psychoneuroimmunology. *Brain, Behavior, and Immunity* (2003) 17(1): 3–12.

Kiecolt-Glaser, J. K., L. McGuire, T. F. Robles, and R. Glaser. Emotions, morbidity, and mortality: New perspectives from psychoneuroimmunology. *Annual Review of Psychology* (2002) 53: 83–107.

Lutgendorf, S. K., and E. S. Costanzo. Psychoneuroimmunology and health psychology: An integrative model. *Brain, Behavior, and Immunity* (2003) 17(4): 225–232.

Lutgendorf, S. K., P. P. Vitaliano, T. Tripp-Reimer, J. H. Harvey, and D. M. Lubaroff. Sense of coherence moderates the relationship between life stress and natural killer cell activity in healthy older adults. *Psychology and Aging* (1999) 14(4): 552–563.

Sense of Control and Positive Health Effects with Aging

Rodin, J. Aging and health: Effects of the sense of control. *Science* (1986) 233(4770): 1271–1276.

Rodin, J. Sense of control: Potentials for intervention. *Annals of the American Academy of Policy and Social Science* (1989) 503: 29–42.

Appendix 2:
Other Useful Resources

AGENCIES, INTERNET, AND BOOKS

Agencies and Internet Resources

AARP
601 E Street, N.W.
Washington, DC 20049
800-424-3410
www.aarp.org
AARP, with a membership in excess of 30 million people aged fifty and older, is a leader in addressing the interests and issues of older people. The organization offers a wealth of information on diverse aspects of aging.

Age Wave
One Embarcadero Center, Suite 3810
San Francisco, CA 94111
415-705-8014
www.agewave.com
Age Wave is a firm created to guide Fortune 500 companies and government groups in product and service development for Baby Boomers and mature adults.

AgingStats.Gov
www.agingstats.gov
This is the Web site of the Federal Interagency Forum on Aging-Related Statistics, which provides a wealth of data on aging.

Alliance for Aging Research
2021 K Street, N.W., Suite 305
Washington, DC 20006
202-293-2856
www.agingresearch.org
The Alliance for Aging Research is America's leading citizen advocacy organization for promoting research in human aging and working to ensure healthy longevity for all Americans.

American Association of Geriatric Psychiatry (AAGP)
7910 Woodmont Avenue
Bethesda, MD 20814
301-654-7850
www.aagpgpa.org
A national professional society dedicated to improving the mental health and well-being of older persons. Informational materials are available for both professionals and the public.

American Federation for Aging Research (AFAR)

70 West 40th Street, 11th Floor
New York, NY 10018
212-703-9977
www.afar.org
AFAR fosters research on the fundamental processes of aging—what scientists call biogerontology—with the goal of extending healthy life and finding cures for diseases that accompany old age. Information on this area is also provided.

American Geriatrics Society

350 Fifth Avenue, Suite 801
New York, NY 10018
212-308-1414
www.americangeriatrics.org
A national professional society dedicated to improving the health and well-being of older persons. Informational materials are available for both professionals and the public.

American Society on Aging (ASA)

833 Market Street, Suite 511
San Francisco, CA 94103
415-974-9600
www.asaging.org
The ASA is a national nonprofit membership organization that informs the public and health professionals about issues that affect quality of life for older people and promotes innovative approaches to meet the needs of older people.

Center on Aging, Health & Humanities
George Washington University
10225 Montgomery Avenue
Kensington, MD 20895
202-895-0230
www.gwumc.edu/cahh
The center houses programs (including the Creativity Discovery Corps) headed by Gene D. Cohen, M.D., Ph.D., at George Washington University in Washington, D.C., with a special focus on studying and promoting creativity and aging.

Center for Elders and Youth in the Arts (CEYA)
3330 Geary Boulevard
San Francisco, CA 94118
415-750-4111
www.ioaging.org/programs/art/art.html
Working with high schools, selected middle schools, local arts agencies, and group sites for the elderly, CEYA teams youths and elders in collaborative educational programming under the instruction of professional visual and performing artists. The center provides an infrastructure for planning, designing, and implementing cross-generational projects and community presentations. Visual and performing artists, playwrights, poets, and musicians have been carefully selected and trained by geriatric professionals and educators to work with elderly and youth.

Appendix 2: Other Useful Resources

Civic Ventures

139 Townsend Street, Suite 505

San Francisco, CA 94107

415-430-0141

www.civicventures.org

Civic Ventures, a think tank as well as an incubator, aims to create ideas and develop programs to help society achieve the greatest return on experience. Its particular focus is on Americans who are redefining the second half of life—people who are not just extending their years on the job but also doing work that adds meaning to these years. Participants want to share their experience while acquiring new experiences. They are inventors, organizers, leaders, activists, teachers, and entrepreneurs who attach deep meaning to the notion of giving back.

Elderhostel

75 Federal Street

Boston, MA 02110-1941

877-426-8056

www.elderhostel.org

Elderhostel describes itself as follows: "Elderhostel is a nonprofit organization providing educational adventures all over the world to adults age 55 and over. Study the literature of Jane Austen in the White Mountains of New Hampshire, or travel to Greece to explore the spectacular art and architecture of its ancient civilization, or conduct field research in Belize to save the endangered dolphin population. Elderhostel is for people on the move who believe learning is a lifelong process."

Elders and Families
Administration on Aging
One Massachusetts Avenue, N.W.
Washington, DC 20201
202-619-0724
www.aoa.gov/eldfam/eldfam.asp
Elders and Families is designed to assist older persons and their care-givers in quickly obtaining information and resources on a variety of aging-related topics. The information is organized to help you become more familiar with issues affecting older adults and the services and opportunities available to assist them.

Elders Share the Arts (ESTA)
138 S. Oxford Street
Brooklyn, NY 11217
718-398-3870
www.elderssharethearts.org
Founded in 1979, ESTA is a nationally recognized arts organization dedicated to bridging generational divides and generating a sense of community through the arts. Its staff of professional artists works with the young and old in underserved communities to transform their life stories into dramatic, literary, and visual presentations that explore social issues, shed light on neighborhood history, and draw from their imaginations answers to community issues and conflicts. The organi-zation also has technical assistance and training materials for developing community-based arts programs.

Experience Corps

2120 L Street, N.W., Suite 610
Washington, DC 20037
202-478-6190
www.experiencecorps.org

Experience Corps offers new adventures in service for Americans over fifty-five. Now in fourteen cities, Experience Corps works to solve serious social problems, beginning with literacy. Today, more than 1,800 Corps members serve as tutors and mentors to children in urban public schools and after-school programs, where they help teach children to read and develop the confidence and skills to succeed in school and in life. Research shows that Experience Corps boosts students' academic performance, helps schools and youth-serving organizations become more successful, strengthens ties between these institutions and surrounding neighborhoods, and enhances the well-being of the volunteers in the process. Experience Corps is a signature program of Civic Ventures.

Fact Sheets, Issue Briefs, and Snapshots

Administration on Aging

One Massachusetts Avenue, N.W.
Washington, DC 20201
202-619-0724
www.aoa.dhhs.gov/press/fact/fact.asp

Fact Sheets, Issue Briefs, and Snapshots are information pieces intended to help prepare everyone for the longevity revolution by heightening awareness about the demographics and challenges of the rapidly advancing twenty-first century. They include an overview of the subject and details related to Administration on Aging initiatives. Also included are various resource listings, such as federal agencies, national organizations, and suggested readings in many languages.

GENCO International, Inc.
PO Box 66
Kensington, MD 20895-0066
301-946-6446
www.GENCO-GAMES.com
GENCO is Gene Cohen's creativity company. It focuses on the development of intergenerational, educational, artistic board games that provide mental exercises for aging.

Generations United
1333 H Street, N.W., Suite 500 W
Washington, DC 20005
202-289-3979
www.gu.org
At this Web site you can obtain information on a list of intergenerational coalitions, links to sites with an intergenerational theme, intergenerational programming, and other intergenerational-related resources.

Gerontological Society of America (GSA)
1030 Fifteenth Street, N.W., Suite 250
Washington, DC 20005-1503
202-842-1275
www.geron.org
The GSA is a national nonprofit professional society that promotes the scientific study of aging in the biological, behavioral, and social sciences. It has a humanities and arts committee with a strong focus on creativity and aging.

Appendix 2: Other Useful Resources

Global Action on Aging (GAA)
PO Box 20022
New York, NY 10025
212-557-3163
www.globalaging.org/index.htm
GAA is a nonprofit organization with special consultative status with the United Nations Economic and Social Council. GAA carries out research on critical emerging topics and publishes the results on its Web site. GAA staff and interns research aging policy and programs in the United States and worldwide. GAA posts materials in all six UN official languages.

International Federation on Ageing (IFA)
4398 Boul. Saint-Laurent, Suite 302
Montreal QC H2W 1Z5 Canada
514-396-3358
www.ifa-fiv.org
The IFA helps link more than 100 associations as well as interested individuals representing or serving older persons in approximately fifty nations around the world.

International Longevity Center (ILC)
60 East 86th Street
New York, NY 10028
212-288-1468
www.ilcusa.org
The ILC is dedicated to the study of longevity internationally, improving the availability of data as well as the identification of programs and approaches to improving the quality of one's added years.

International Psychogeriatric Association (IPA)
550 Frontage Road, Suite 2820
Northfield, IL 60093
847-501-3310
www.ipa-online.org
The IPA is an international professional society dedicated to improving the mental health and well-being of older persons. Informational materials are available for both professionals and the public.

National Center for Creative Aging (NCCA)
138 S. Oxford Street
Brooklyn, NY 11217
718-398-3870
www.creativeaging.org
The National Center for Creative Aging (NCCA) is dedicated to fostering an understanding of the vital relationship between creative expression and the quality of life of older people through serving as a clearinghouse for the exchange of information regarding creativity and aging; evaluating arts and aging programs in order to identify best-practice models and promote their replication nationally; advocating public policy in support of quality arts programs as essential to the well-being of older people; encouraging efforts to document the role of creative expression in the lives of older people; providing quality training and education; and developing and disseminating resource materials concerning creative aging.

Appendix 2: *Other Useful Resources*

National Council on the Aging (NCOA)
300 D Street, S.W., Suite 801
Washington, DC 20024
202-479-1200
www.ncoa.org
The NCOA is an association of organizations and individuals dedicated to promoting the self-determination, well-being, and continuing contributions of older persons through service, education, and advocacy. Its members include professionals and volunteers, service providers, consumer groups, businesses, government agencies, religious groups, and voluntary organizations.

National Institute on Aging (NIA)
31 Center Drive, MSC 2292
Building 31, Room 5C27
Bethesda, MD 20892
800-222-2225
www.nih.gov/nia
The NIA is the federal research program most involved in supporting studies of aging. In addition to providing information on research findings, much practical information is offered through the institute's diverse publications—especially their Age Pages.

North Carolina Center for Creative Retirement
Reuter Center, CPO #5000
University of North Carolina at Asheville (NCCCR)
One University Heights
Asheville, NC 28804-8516
828-251-6140
www.unca.edu/ncccr
The NCCCR has the threefold purpose of promoting lifelong learning, leadership, and community service opportunities for retirement-aged individuals. Most of NCCCR programs are in the Asheville area, but some are carried out in collaboration with other organizations in other parts of North Carolina or across the country.

Senior Corps
1201 New York Avenue, N.W.
Washington, DC 20525
800-424-8867
www.seniorcorps.org
Through the Senior Corps, nearly half a million Americans age fifty-five and older share their time and talents to help their communities. The Corps's programs include the Foster Grandparent Program, RSVP (Retired Seniors Volunteer Program), and Senior Companion Program.

Senior Leadership Montgomery
5705 Arundel Avenue, Suite 200
Rockville, MD 20852
301-881-3333
www.leadershipmontgomerymd.org/senior.asp
Senior Leadership Montgomery is for people fifty-five and older who are retired, semiretired, or about to retire and are ready to share their

wisdom and talent with like-minded individuals who want to make a difference. Approximately twenty-five participants representing their culturally, ethnically, socially, economically, and geographically diverse communities are selected. The selection committee tries to identify individuals who will use their leadership for the benefit of the community. The program provides a dynamic learning experience highlighted by community action projects carried out by the participants over the course of the program. During the sessions participants learn firsthand about issues and needs of the community and meet inspiring community leaders. Sessions include community sampler tours.

SeniorNet
1171 Homestead Road, Suite 280
Santa Clara, CA 95050
408-615-0699
www.seniornet.org

Small Business Administration (SBA)
409 Third Street, S.W.
Washington, DC 20416
800-827-5722
www.sba.gov
The U.S. Small Business Administration provides financial, technical, and managerial assistance to help Americans start, run, and grow their businesses. With a portfolio of more than $50 billion, SBA is the nation's largest single backer of small businesses.

SPRY Foundation
10 G Street, N.E., Suite 600
Washington, DC 20002
202-216-0401
www.spry.org
The SPRY (Setting Priorities for Retirement Years) Foundation is an independent nonprofit research and education organization that helps people prepare for successful aging. SPRY emphasizes planning and prevention-oriented strategies in four key areas: health and wellness, mental health, financial security, and life engagement.

ThirdAge Inc.
25 Stillman Street, Suite 102
San Francisco, CA 94107-1309
www.thirdage.com
ThirdAge is an online media and direct-marketing company focused exclusively on serving the needs of midlife adults—generally those in their forties, fifties, and sixties—and others who want to build a genuine relationship with them.

U.S. Census Bureau
(For information on aging)
Hagerstown Telephone Center
1125 Opal Court, 2nd Floor
Hagerstown, MD 21740
800-321-1995
www.census.gov/population/www/socdemo/age.html
The Age Data section provides very well done, brief, and in-depth data reports on both Baby Boomers and older adults.

U.S. Senate Special Committee on Aging

Senator Gordon H. Smith, Chairman
G31 Dirksen Senate Office Building
Washington, DC 20510
202-224-5364
aging.senate.gov/public/index.cfm?Fuseaction=Home.Home
The Special Committee on Aging has served as a focal point in the
U.S. Senate for discussion and debate on matters relating to older
Americans. Often the committee submits its findings and recommendations for legislation to the Senate. In addition, the committee
publishes materials to assist those interested in public policies related
to older adults.

Appendix 2: Other Useful Resources

Books

General Books on Aging
Binstock, Robert H., and Linda K. George, editors. 2001. *Handbook of Aging and the Social Sciences*, 5th edition. New York: Academic Press.

Birren, James E., editor-in-chief. 2001. *The Handbooks of Aging*. New York: Academic Press.

Birren, James E., and K. Warner Schaie, editors. 2001. *Handbook of the Psychology of Aging*, 5th edition. New York: Academic Press.

Cohen, Gene D. 2001. *The Creative Age: Awakening Human Potential in the Second Half of Life*. New York: HarperCollins Publishers.

Maddox, George, editor. 2001. *The Encyclopedia of Aging,* 3rd edition. New York: Springer Publishing Company.

Masoro, Edward J., and Steven N. Austed. 2001. *Handbook of the Biology of Aging,* 5th edition. New York: Academic Press.

Ricklefs, Robert E., and Caleb E. Finch. 1995. *Aging: A Natural History.* New York: Scientific American Library.

Rowe, John W., and Robert L. Kahn. 1998. *Successful Aging.* New York: Pantheon Books.

Books on or Relevant to Understanding the Aging Brain and Mind
Cohen, Gene D. 1988. *The Brain in Human Aging.* New York: Springer Publishing Company.

Cohen, Gene D. 2001. *The Creative Age: Awakening Human Potential in the Second Half of Life.* New York: HarperCollins Publishers.

Goldberg, Elkhonon. 2005. *The Wisdom Paradox*. New York: Gotham Books.

Park, Denise, and Norbert Schwarz, editors. 2000. *Cognitive Aging*. Philadelphia: Psychology Press.

Restak, Richard. 2001. *The Secret Life of the Brain*. Washington, DC: Joseph Henry Press.

Schwartz, Jeffrey M., and Sharon Begley. 2002. *The Mind and the Brain*. New York: Regan Books.

Stern, Paul C., and Laura L. Carstensen, editors. 2000. *The Aging Mind*. Washington, DC: National Academy Press.

Books Addressing Psychological Development with Aging
Cohen, Gene D. 2001. *The Creative Age: Awakening Human Potential in the Second Half of Life*. New York: HarperCollins Publishers.

Colarusso, C. A., and Robert A. Nemiroff. 1981. *Adult Development*. New York: Plenum Press.

Erikson, Erik E. 1980. *Identity and the Life Cycle*. New York: W. W. Norton.

Erikson, Erik E. 1997. *The Life Cycle Completed*. New York: W. W. Norton.

Gould, Roger L. 1978. *Transformations: Growth and Change in Adult Life*. New York: Touchstone Books.

Lachman, Margie E., editor. 2001. *Handbook of Midlife Development*. New York: John Wiley & Sons.

Levinson, Daniel J. 1978. *The Seasons of a Man's Life*. New York: Ballantine Books.

Pollock, George H., and Stanley I. Greenspan, editors. 1993. *The Course of Life,* volume 6, *Late Adulthood.* Madison, CT: International Universities Press.

Pollock, George H., and Stanley I. Greenspan, editors. 1997. *The Course of Life,* volume 7, *Completing the Journey.* Madison, CT: International Universities Press.

Sheehy, Gail. 1974. *Passages.* New York: Bantam Books.

Sheehy, Gail. 1995. *New Passages.* New York: Ballantine Books.

Vaillant, George. 2002. *Aging Well.* New York: Little, Brown and Company.

Books on the Retirement Period of Life
Brock, Fred. 2004. *Retire on Less Than You Think: The New York Times Guide to Planning Your Financial Future.* New York: Times Books.

Cohen, Gene D. 2001. *The Creative Age: Awakening Human Potential in the Second Half of Life.* New York: HarperCollins Publishers.

Freedman, Marc. 1999. *Prime Time: How Baby Boomers Will Revolutionize Retirement and Transform America.* New York: Public Affairs.

Hinden, Stan. 2001. *How to Retire Happy: Everything You Need to Know About the Twelve Most Important Decisions You Must Make Before You Retire.* New York: McGraw-Hill.

Howells, John. 2000. *Retirement on a Shoestring,* 3rd edition. Guilford, CT: Globe Pequot Press.

Jacobs, Ruth Harriet. 1993. *Be an Outrageous Older Woman.* Manchester, CT: KIT Publisher.

Kerschner, Helen K., and John E. Hansan, editors. 1996. *365 Ways . . . Retirees' Resource Guide for Productive Lifestyles.* Westport, CT: Greenwood Press.

Lindeman, Bard. 1998. *Be an Outrageous Older Man.* Manchester, CT: KIT Publisher.

Trafford, Abigail. 2004. *My Time.* New York: Basic Books.

Vandervelde, Maryanne. 2004. *Retirement for Two.* New York: Bantam Books.

Special Section: Books for Children in Which Aging and Older Adults Are Portrayed in a Positive Light

Background by Gene Cohen

Numerous studies have found that although American children have a positive view of older adults in their own family, they have a negative overall view of aging. No one had been able to explain this paradox. I thought long and hard about this and about what, in addition to children's parents and family, would have an early influence on their view of aging. It clicked for me when my wife and I took our daughter, Eliana, at age four to see the movie *101 Dalmatians*, in which Glenn Close portrays the older, evil Cruella DeVil. After the movie, we read the original book, *101 Dalmatians*, to Eliana as we put her to bed. Later that night, around three o'clock in the morning, Eliana marched into our bedroom and stamped her feet on the floor yelling, "I never want Cruella in our house—*not even in a book!*" I then postulated that a major factor contributing to children's negative attitudes about aging is that the earliest literature given to young children, starting with the fairy tales, typically portray older people as wicked, weird, or weak. Consider Cinderella's malevolent stepmother, the evil old witch in Hansel and Gretel, the scheming Rumpelstiltskin, and the child-abusing old lady who lived in a shoe.

These are great classics, but what if families want to offer their children literature with positive images of aging? To begin with, very few reading lists are available that suggest books for the very young in which older adults are portrayed positively, and none have been easily accessible and widely recognized. I began asking librarians if they knew of any, and none that I talked to did, although they all thought that such a list would be of great value and interest. After further research on how children's books portray older characters, I found that there was very little character development, apart from the usual—wicked, weird, or weak. One study reported that the most common way an older character would be introduced is to be described in one word—"old"—as if that said it all. Even when older people aren't portrayed negatively, they are rarely portrayed as protagonists in children's literature. Psychiatrist and fairy tale expert Allan Chinen reported a cross-cultural study he conducted on 2,500 fairy tales randomly selected from the Berkeley University library. He found that "only 2 percent featured a protagonist who was identified as old."

I realized, too, that the writers could not be criticized, because, after all, look at how little science has contributed historically to understanding the development of older persons in real life. The illusion of knowledge about aging led science, society, and writers of children's books to view older people as offering very little positive or interesting to elaborate on.

Appendix 2: Other Useful Resources

To the extent that the youth of America have a better sense of the potential that can accompany aging, they can develop a life cycle perspective that improves their preparedness for both the problems and the possibilities of later life as well as enhancing their attitudes toward positive intergenerational relationships. Too many Americans, inundated as they are by negative myths and stereotypes about aging, deny aging and consequently fail to adequately prepare for it.

To address this gap, I approached the Association for Library Service to Children (ALSC) of the American Library Association to explore a collaboration to create an authoritative, annotated, up-to-date, widely disseminated reading list for children, describing books that portray aging and older adults in a positive light. The effort was one of the initiatives that came out of my center's newly established public education program on aging—known as the SEA Change Program, or Societal Education about Aging for Change. The ALSC enthusiastically agreed to collaborate and used their infrastructure of librarians across the country to assemble an annotated list of ninety-one books presenting aging realistically and positively for children from prekindergarten through sixth grade, indicating reading levels. The Center on Aging, Health & Humanities had the list evaluated by more than 100 librarians and 100 consumers from around the country, and the evaluations came back overwhelmingly favorable.

The reading list is included in its entirety at www.gwumc.edu/cahh/booklist. This Web site also provides an interactive option for readers to recommend additional books of this nature.

Index

Index

Index

DeMarco, Sally, 43–45
Dementia, 26. *See also* Alzheimer's
  disease
Dempsey, William T. J. "Bill," 82
Dendrites, 69, 84, 101
Depression, 4, 12, 14, 28, 46, 58, 61,
  86, 123, 126, 129, 175, 178
Developmental intelligence, xix–xx,
  34–39, 45, 48, 51, 55, 62, 64, 66,
  96, 97, 123, 173
  as basis for wisdom, 52, 95, 103
  and brain changes, 101
  defined, 35
  integration of components of, 35
  role of emotions in, 45–46
  and social intelligence, 116, 120
  See also Psychological development
Diamond, Marion, 6
Dickens, Charles, 129
Disengagement theory, 117–118, 147
Dreams, 64, 65, 129, 170
  dream journals, 161–162
Driscoll, Denise, 170
Drives, 34, 48, 65, 92. *See also* Inner
  Push
Drucker, Peter, 51
Dualistic thinking, 37, 38, 96
Duke University Center for Cognitive
  Neuroscience, 20
Dunton, James, 64–65, 162

Education, xxi, 153, 160. *See also*
  Learning
Einstein, Albert, 11, 29, 43, 169, 170
Ekerdt, David J., 137–138
Elders Share the Arts (Brooklyn),
  177–178
E-mail, 161. *See also* Internet
Emotion(s), 4, 9, 13–18, 28, 30, 39, 57,
  64, 84, 120, 122, 123
  controlling, 12, 16–17, 18, 46, 62
  and cortex/limbic system
    connectivity, 16–17
  emotional intelligence, 35, 46, 95

integrated with reason, 97, 98, 99
  and midlife reevaluation, 61
  negative, 15–16, 18, 45–46, 85, 131
  positive, 14–15, 18, 27, 46, 53, 85
  role in developmental intelligence,
    45–46
Empowerment, 137, 143, 157, 179
Encore developmental phase, xvii, xix,
  53, 82–89
  and creativity, 173
  and postformal thinking, 100–101
  and social intelligence, 120–121
Endorphins, 25
Erikson, Erik, xvi–xvii, 41–42, 51
Exercise, 5
  mental, 24, 26, 110, 159
  physical, 12, 24–25, 26, 149–150

Falstaff (Verdi), 81
Fear, 16, 17, 18, 58, 106
Financial planning, 144, 156
Frank, Elinor, 100–101
Franklin, Anna, 90–91
Franklin, Benjamin, 93, 105
Freud, Sigmund, xvi, 41, 42
Friendships, 149–150, 157, 164, 181
Frith, Christopher D., 77

Gage, Fred, 12–13
Galdikas Biruté Mary, 136, 137
Galileo, 73
Gardening, 170
Gardner, Howard, 169–170
Gender, 122–125
Genealogy, 163
Genetic issues, 9
George Washington University, Center
  on Aging, Health, and Humanities,
  55
Georgia State University, 84
Gerontology, 55
Giving back, xviii, 76, 78, 150–152. *See
  also* Volunteerism
Glasgow, Ellen, 165

Index

Index

Index